CHINESE ASTROLOGY AND ASTRONOMY

An Outside History

CHINESE ASTROLOGY AND ASTRONOMY

An Outside History

Xiaoyuan Jiang
Shanghai Jiao Tong University, China

Translated by

Wenan Chen
Ningbo University, China

NEW JERSEY · LONDON · SINGAPORE · BEIJING · SHANGHAI · HONG KONG · TAIPEI · CHENNAI · TOKYO

Published by

World Scientific Publishing Co. Pte. Ltd.

5 Toh Tuck Link, Singapore 596224

USA office: 27 Warren Street, Suite 401-402, Hackensack, NJ 07601

UK office: 57 Shelton Street, Covent Garden, London WC2H 9HE

Library of Congress Control Number: 2020951935

British Library Cataloguing-in-Publication Data
A catalogue record for this book is available from the British Library.

This book is published with the financial support of Chinese Fund for Humanities and Social Sciences.

《天学外史》
Originally published in Chinese by The Shanghai Jiao Tong University Press
Copyright © The Shanghai Jiao Tong University Press 2015

CHINESE ASTROLOGY AND ASTRONOMY
An Outside History

ISBN 978-981-122-345-7 (hardcover)
ISBN 978-981-122-346-4 (ebook for institutions)
ISBN 978-981-122-347-1 (ebook for individuals)

For any available supplementary material, please visit
https://www.worldscientific.com/worldscibooks/10.1142/11914#t=suppl

Typeset by Stallion Press
Email: enquiries@stallionpress.com

Printed in Singapore

Preface

Seven years ago, when my old friend Jiang Xiaoyuan finished his *Tianxue Zhenyuan (The Truth of the Sciences of the Heaven)*, I was kindly invited by him to write a preface for it. In the passage I wrote then, I digressed a lot from what was in the book to dwell on the history of science and the Whig interpretation of history, based on the current study of the history of science in China; but actually, the Whig history is, in a way, quite closely related to the theme of *Tianxue Zhenyuan*. The book has received a favorable reception after its publication; even now, seven years later, it still occupies a unique position among the numerous monographs on the history of ancient Chinese science and sees many scholarly reviews one after the other featuring either great compliments or fierce criticism. Of course, such a huge success is to be expected, given the author's profound academic knowledge, the novel perspective, and the abundant historical materials of reference.

Seven years later, when *Tianxue Waishi (An Outside History of Chinese Astronomy)*, the companion piece to *Tianxue Zhenyuan*, was completed and about to go to press, my friend again asked me to preface it. On the one hand, though I still felt incompetent to take on this task and improper to see it as a soapbox for whatever topic I'd like to dwell on, emboldened by the kind favor of my old friend, I thought I might as well accept the invitation since I had already "ventured" to "sing a different tune" seven years ago, so my new preface might add just a little bit more gravity to my "crime". On the other hand, though I am a layman in the realm of the history of astronomy, especially the history of ancient Chinese astronomy, I am very interested in the latest developments in this field and have some ideas of my own about the historiography of science. So I found it the perfect opportunity to share in my preface these ideas and some other thoughts, which, random as they might be, at least, are all true words from my heart — though even now it is still hard to "tell the truth".

Zhang Xiaoyuan named his new book "Tianxue Waishi", literally "An Outside History of Chinese Astronomy". From this title, we can already glimpse the innate attitude of the author. The reason he used the term "tianxue" instead of "tianwenxue (astronomy)" has been detailed in both this book (Chapter 2) and his former book *Tianxue Zhenyuan*: generally, it is to distinguish the various ancient Chinese theories about "astronomy" from the now popularized modern astronomy of Western origin. This is of great significance and clarifies the author's stance. As for the term "waishi (outside history)", this shows expressly the author's research focus.

I have done some research on the basic theoretical issues of the history of science, or the historiography of science, so I have used quite a lot of times the term "outside history" from the Western history of science and philosophy of science in my essays and books. Domestic scholars in the two fields often translated the term into "waishi" as opposed to "inside history", or "neishi" in Chinese. However, I remember that several years ago, during a conversation with Mr. Ge Ge, a senior scholar in the history of physics in China, he had noted that such a translation did not accord with the original concept of the term "neishi", because in Chinese history, "waishi" was used as opposed to "zhengshi (official history)", its meaning much closer to that of "pseudo-history". A similar example is the oft-used term "tongshi (general history)" in the domestic history of science. Historically in China, it has denoted the opposite of the term "duandaishi (dynastic history)", rather than an overarching history that encompasses all the disciplines as opposed to "xuekeshi (disciplinary history)" in today's history of science. Hence, in light of the current accepted meaning of "xuekeshi", it is better to put at its opposite end the term "zongheshi (comprehensive history)" instead of "tongshi". This, of course, has already become a topic as to how to deal with the existing terms in Chinese history in the translation of modern Western concepts related to the history of science.

Due to the complexities in the translation and application of concepts, my friend expounded in the first chapter of his new book the three-fold meanings of the important concept "waishi" according to his own understanding. So, the first chapter can also be regarded as a unique article on the historiography of science that combines with it personal practical experience — a practice uncommonly seen in the publications of researchers in the history of science.

Perhaps, it is exactly my friend's habitual contemplation on the history of science from a theoretical perspective that imparts his research a unique feature. This book, *Tianxue Waishi*, while inheriting the merits of *Tianxue Zhenyuan*, has gone a step further to make new explorations on many issues, coming up with a slew of bold yet well-founded arguments, including many that pose a challenge to the authoritative opinions. Among them, what interests me most is his investigation into the nature and function of the ancient Chinese "tianxue" *(the sciences of the heaven)* and its difference from the "astronomy" we commonly refer to now, namely, the modern Western astronomy. Of course, conclusions from such investigations are very likely to touch a nerve in those self-proclaimed "patriots" who put inordinate weight on the "scientific achievements" of ancient China and would go to great Whiggish extremes to prove "Chinese leadership" in all disciplines and important scientific issues.

I am not groundless here. In the first chapter of this book, my friend talked about his report entitled *"Patriotic Education Should Not Be the Goal of the Study of the History of Science"* at a national seminar held in Yantai, Shandong Province, in 1986, and the heated debate it had triggered; he commented that his view "may seem aggressive at the time, but it is now rather acceptable for most of the scholars". However, personally speaking, the truth is, not really. Just recently, there was a bitter dispute over the existence of science in ancient China in the press, which makes it very clear that my friend's view is still met with many disapproving voices and strong opposition even now.

In the recent discussions on whether ancient China had seen true science, many of the proponents had actually all turned a blind eye to the different meanings of the word "science" in different contexts.

The word "science" has different meanings in both Chinese and English. Most commonly, it refers to modern science born in Europe. In other uses, technology is also included, or, by extension, even any correct and effective methodologies, ideas, etc. The meaning of the word "science" applied when we say, for example, the history of science in the Song Dynasty of China, the history of science in ancient India, or the history of science in ancient Greece, is apparently not modern science in its most common sense, since the "science" of ancient China or India is a completely different matter, even though Ancient Greek traditions do share the same origin with the modern European science. As a matter of fact, one of the important characteristics of modern European science is a systematic knowledge of nature. Just as a scholar in our country has long argued, ancient China did not have physics, only the knowledge of physics. The reason lies precisely in the absence of a systematic knowledge of nature. This, of course, does not prevent us from still using "the history of ancient Chinese physics" to refer to the study of the ancient Chinese knowledge in physics. So it goes with the history of ancient Chinese astronomy. *Tianxue Zhenyuan* and *Tianxue Waishi* both make a point of using the term "tianxue" (the sciences of the heaven) rather than "astronomy" in the study of "the history of ancient Chinese astronomy (if we can still say so)"; in doing so, they are exactly trying to avoid the confusion of concepts that may arise from using the same term to refer to different things.

Actually, in that dispute over science, many vociferous proponents of the belief that there was science in ancient China were driven by a more profound rationale. For example, one of them explained with clarity that: "Nowadays, a considerable number of Chinese scientists and engineers, especially of the younger generation, fed with 'Eurocentric' education in science from an early age, know little, or even far less than little, about our great traditional Chinese culture. So, a pressing matter of the moment is to increase awareness and build national confidence, not to 'indulge ourselves' in the achievements of our ancestors". To attain such a goal, I am afraid that the current textbooks for science teaching in Chinese primary and secondary schools — not to mention universities — will have to be overhauled, for an obvious reason: how much of the content has a Chinese origin? And how much of the content belongs to the ancient "science" of China? Today, the slogan "invigorating the country through science, technology and education" is on every man's tongue; yet, can we really actualize it by relying on the ancient "science" of China that has no connection with modern science and building national confidence based thereon? The answer is a categorical no.

With a clear knowledge of all the relevant concepts, the question whether ancient China had science is not really hard to answer, as it were. At least, as to whether ancient China had astronomy, *Tianxue Waishi* (and *Tianxue Zhenyuan*) has given a clear answer.

This passage just shares some of my thoughts that were stirred up after reading the manuscript of *Tianxue Waishi*, which, of course, covers much more than that; but after all, a preface cannot possibly be so all-encompassing that it touches upon every point in the book. Furthermore, a preface can only represent the personal opinion of the preface writer, and feedback of the true significance of the text should come from the readers. After a book is published, the task of interpretation is left to the readers. So even the author himself, let alone the preface writer, can only observe from the sidelines the readers' reception. But I believe any truly discerning mind is bound to find something valuable and instructive in this book.

Liu Bing
Tiantandongli, Beijing
September 6, 1998

Author's Note for the
New Edition

This book was first published in 1999, eight years after my *Tianxue Zhenyuan* came out. Thanks to the generous compliments from readers, I received the National Book Award for *Tianxue Zhenyuan* a year later — by then, it had already been reprinted and reissued multiple times, and even seen a traditional-Chinese-character edition published in Taiwan (1995). A friend of mine (Prof. Tian Song, now teaching at Beijing Normal University) praised it by saying that it was "as interesting as a detective story", while another (the late Prof. Ge Ge) thought it "not exoteric enough". During those few years, I had made some new investigations into the ancient Chinese *tianxue,* which gave me new reflections. And by coincidence, the Shanghai People's Publishing House came to me, asking me to write a new book at the time. That led to the companion piece *Tianxue Waishi* to *Tianxue Zhenyuan*, the two complementing and reinforcing each other.

When I was writing the book *Tianxue Waishi*, the criticism "not exoteric enough" kept ringing in my ears, so I paid extra attention to keeping the language simple to enable understanding by the readers, hoping that this would have more appeal to the common multitude. However, as to the result, I really have no idea. Though I have made quite a few "exoteric" efforts, I still adhered to the basics of academic writings, providing the origin for all the important historical evidence and texts.

To my surprise, three years after its publication, in 2002, this book was awarded "Excellent Book" by "The First Taiwan Wu Dayou Popular Science Works Prize" together with other four original mainland works. I had never thought that I had made any contribution to the popularization of science and I also had not intended the book as a popular science book, yet I received such an unexpected honor! Either way, it seems that my "exoteric" efforts are well recognized.

The new edition does not make any alteration to the content, but its typographical design is improved to offer a better reading experience.

Jiang Xiaoyuan
School of History and Culture of Science,
Shanghai Jiao Tong University
April 28, 2016

Foreword

Witticism about equation

In 1909, Danish philosopher *Anton Thomsen* (1877–1915), then the brother-in-law of the great physicist Niels Bohr, wrote an ebullient letter of thanks to Bohr after receiving a paper on physics from him. His letter begins as follows:

"Dear Niels,
Many thanks for your paper. I planned to read it until I came across the first equation in the paper, but unfortunately, it appears in Page 2".[1]

Thomsen did what he said, and did not continue reading it.

Eighty years later, another great physicist Stephen Hawking, who was at the peak of his fame, uttered the following words in a speech delivered in October 1989.

Normally it needs equations to learn science. Though equations are concise and precise methods and means to describe mathematical ideas, most people stay at a respectful distance from them. Recently, I was writing a popular book. Someone told me that one equation I put in the book would cut the sales volume by half. I did include one equation, which is

---

[1] Cited from D. Favrholdt's *The Philosophical Background of Niles Bohr*, Translated by Ge Ge, Science Press, (1993), p. 88.

the famous equation of Einstein $E = mc^2$. Perhaps without this equation, the sales volume of my book could have been doubled.[2]

It can be seen that equations are considered such a nuisance, both in foreign countries and in China.

As a matter of fact, today's readers all have received training in equations as long as they have finished their secondary education. Unfortunately, most people have returned the knowledge about equations to their teachers after they graduate. Teachers of mathematicians must have felt sorry at the thought of it.

Seeing such interesting stories about equation, you must have had the hunch that there will not be any equations in this book. The sales volume cannot be big, and I simply do not want that to be halved.

Companion volume

This book is a sister of *Tianxue Zhenyuan (The Truth of the Sciences of the Heaven)*. But the two sisters differ a little bit from each other in attire.

Though there are no equations in *Tianxue Zhenyuan,* in form it appears too rigid and serious.[3] I plan to try something new in form with the book *Tianxue Waishi (An Outside History of Chinese Astronomy)*. This book will no longer make use of three-tiered sub-titles, but use the format of numbered chapters and sections. The synopses of these sections shall appear in turn in the table of contents. The thread of narration among various sections is consistent.

This book *Tianxue Waishi* is an extension of the theme of *Tianxue Zhenyuan (The Truth of the Sciences of the Heaven)*. Since it is a companion piece, it is complementary to *Tianxue Zhenyuan*. It has been eight years since *Tianxue Zhenyuan* came out. For eight years, fellow researchers, my graduate students, and I, and especially those young brilliant minds, have achieved many exciting new results, which are naturally discussed in this book.

---

[2] Stephen W. Hawking: "Hawking's Lectures", Hunan Science and Technology Press, 1994 edition, p. 21.

[3] Some find that *Tianxue Zhenyuan* is fascinating. For instance, Zhonghua Dushu Bao (China Reading News) dated March 11, 1998 had an article by Du Yan entitled "Feeling the Pulse of God", saying: "Among the books I've read in recent years, I find three academic books are as intriguing as detective stories. The first is Ye Shuxian's *Philosophy of Chinese Mythology*; the second is Jiang Xiaoyuan's *Tianxue Zhenyuan* and the third is ...."

Contents

Chapter 1

Introduction

1. What is *waishi*? — Three meanings

The word *waishi* (*outside history*) in the title of this book has three meanings.

First, as used in ancient China, *waishi* (outside history) means "unofficial history", contrary to the so-called "official history" (*zhengshi*). Take Emperor Wu (156 B.C.–87 B.C.) of the Western Han Dynasty (202 B.C.–8 A.D.) as an example, *Wudi Ji* (*Annals of Emperor Wu*), the 6th volume of *Han Shu* (*Book of Han*) provides his official history, while the Wei-Jin-era fable *Hanwu Gushi* (*Stories of Emperor Wu of Han*) is an unofficial history. One of the most well-known unofficial history stories in China is *Rulin Waishi* (*The Scholars*), authored by Wu Jingzi during the Qing Dynasty (1636–1912), and this further added to the popularity of *waishi* as a book title. *Zhongguo Tianwenxue Shi* (*The History of Chinese Astronomy*) (1955) authored by Chen Zungui and *The History of Chinese Astronomy* (1981) compiled by a research team on the history of Chinese astronomy led by Xi Zezong (1927–2009) and Bo Shuren (1934–1997) form part of the *Annals of Emperor Wu* in the field of history of Chinese astronomy. Having been published a long time ago and enjoying great reputation, these two official histories are too exceptional to be dwarfed (but they probably should be revised). However, unofficial histories of Chinese astronomy are scarce.

Second, I have given this word *waishi* the meaning "history concerning foreign countries". In ancient China, many works touched upon the similarities and differences between the astronomy in China and that in foreign countries. But the aforementioned two books about the history of Chinese astronomy did not include much about this topic (because there were few studies on this topic at that time). *Tianxue Zhenyuan*, written by me and published in 1991, contributes one chapter — the longest chapter though — to the discussion of the similarities between Chinese and foreign astronomy in ancient times. This book will give more information on this topic. Actually, the history of communication between the Chinese and the study of astronomy is an indispensable part of the history of Chinese astronomy.

Third, waishi (*outside history*) indicates that my study of Chinese astronomy history is interdisciplinary, not intradisciplinary. This is the most academic indication that *waishi* has. Intradisciplinary studies occur within the scope of an academic discipline, including all figures, events, achievements, methods, and works of this discipline in different eras. *The History of Chinese Astronomy* is such an intradisciplinary work. But interdisciplinary studies give more emphasis to the interaction and interplay of the discipline with its external environment as well as its social function. The term external environment includes geography, cultures, customs, political environment, and military environment. One of my earlier works, *Tianxue Zhenyuan*, was the first interdisciplinary work on the history of Chinese astronomy.

2. History of science — A branch of history

The history of science is the study of the development of science and scientific knowledge, which can be traced back many centuries, but didn't form with any real sense until the 20th century. Today in China, researchers on the history of science belong to the huge family of science researchers. For example, the Institute of History of Natural Science, the authority on the study of the history of science in China to which the Chinese Society of History of Science and Technology of China is affiliated, is under the jurisdiction of the Chinese Academy of Sciences (CAS). Researchers on the subjects of history of mathematics, physics, and chemistry belong to universities in China in Departments of Mathematics,

Physics, and Chemistry. I myself, as a researcher on the history of astronomy, serve in the Shanghai Astronomical Observatory, also a part of CAS. However, the picture is quite different in foreign countries, where researchers working on the history of different science in universities belong to the Department of History. Although the study of history of science requires specialized scientific knowledge — which makes these scholars more like scientists than historians — the history of science is, in essence, "history". Therefore, it is reasonable and beneficial to regard the history of science as a branch of historical studies.

3. Theoretical issues in history

It's better to keep the dream of getting to the "real history"

Thus, we cannot avoid discussing some basic theoretical issues of history. Gone are the days when people were debating what the right method of studying history was, summarizing an argument from historical events, or raising an argument first and then selecting supporting historical sources. In recent decades, various foreign theories of history have been introduced in China, such as the idea of "real history", history that is faithful to the past. At first glance, we may subconsciously take this concept for granted and think that we should make it a goal to strive for, but we may neglect the fact that it is extremely difficult for us to reach the "real history". I have heard that in the United States, any university teacher claiming to have learned the "real history", might be disqualified from teaching due to his knowledge of outdated and obsolete theory. This may be an exaggeration, but if we give it a second thought, we will find that getting to know the "real history" has indeed become a beautiful dream that is hard to realize. Now here comes the question: since chances are slim to realize this dream, should we just give it up? According to me, personally, I think we should keep it — after all, humans cannot live without dreams.

Such issues were not touched upon by Chinese researchers on history of science, and were even more rarely discussed by historians in the past. Many scholars, including me, years ago, when writing their papers, were convinced they were presenting the "real history". Of course, there were scholars who thought differently. For example, Professor Li Zhichao, an active-minded scholar, wrote the following words in his recent paper[1]:

It is a fact the study of history of science has not always been objective... We must strive to make the history of science a "real history", which is really what I'm thinking about. But it's impossible to achieve this goal overnight; there is a long way to go. At present, it is good enough that this goal is accepted by all of us.

History is not insentient or inexorable. We should not be irrational or insentient in judging historical events, may it be praise or criticism. History in general studies both "good" and "evil" historical events and both successes and failures, and it may receive praises or criticism. However, historical events recorded in the history of science are mainly good and successful ones that are praiseworthy. At least, the history of science in China should set up a good example for posterity and be passed on as teaching materials for moral education. The collection of historical sources is not the ultimate goal of the history of science. The sources collected should be used for education. As long as they are well protected, there is no need to absolutely reject fictional and literary interpretations.

In his words, the "real history" seems beyond our reach. His parody of Confucius's words — "History is not insentient or inexorable". — is indeed great, from certain perspectives. However, the matter of regarding the history of science as teaching material for moral education may be opposed by many researchers today. However, this opinion may have deeper implications. With regard to this issue, we can take a look at what Professor Gu Jiegang (1893–1980), a well-known Chinese historian, told nearly seventy years ago:

> *Whether an event is good or evil has nothing to do with us. Our duty is to describe the event. If the politicians want to carry forward the spirit of a nation or the educators want to improve the social climate, they can select sources from us and use them in the way they want.*[2]

---

[1] Li Zhichao. *Tian Ren Gu Yi: Zhongguo Kexueshi Lungang* (*On History of Chinese Science*). Zhengzhou: Henan Education Press, (1995), p. 9.

[2] Gu Jiegang. the prologue of *Mi Shi* (*The History of Riddles*). See Qian Nanyang. *Mi Shi* (*The History of Riddles*). Shanghai: Shanghai Art and Literature Publishing House, (1986), p. 8.

"Striving to make the history of science a 'real history'" and "describing the event" share the same implications. Besides, not all historical events that are dealt with in the study of the history of science are good and successful.

4. Patriotic education: Yes or no in the history of science

This reminds me of my experience at a seminar on the history of science held at Yantai, Shandong, in 1986. I ventured to make a report titled *"Patriotic Education Should Not Be the Goal of the Study of the History of Science"*, and in this I stated my idea that if we set up the theme and goal of facilitating patriotic education of the people first, the truth-seeking process of the study of the history of science would be impeded — admittedly, I was a faithful supporter of the "real history" at the time.[3] The report sparked off such a heated debate at the meeting that the president had to repeatedly ask the participants not to confine discussions to this topic alone. The scholars present were divided into two factions — most of those that opposed my opinion and reserved their judgment were senior scholars; most of the young scholars enthusiastically supported and bravely defended my view. Such a view might have seemed aggressive at the time, but it is now quite acceptable for most of the scholars. In fact, my view is in line with what Gu Jiegang said seventy years ago.

When studying the history of science, the more one thinks about these basic theoretical issues, the more one feels troubled. If one were convinced that the "real history" was within reach, one would be very confident in his studies; or, if one, at the very beginning, took moral education as the goal of the studies (and if he believed it was impossible to reach the "real history"), one would not feel sorry, despite him not being able to provide "real history".

However, if one thought that it was impossible to get to the "real history", but at the same time was reluctant to make moral education the goal of his studies, how should one continue with studies then?

---

[3] This report was later published in *Discovery of Nature*, Vol. 5. No. 4 (1986).

5. Chronicling, concept analysis, and sociological approach

Actually, there is no need to be too worried about this as there are solutions. The study of the history of science, like other academic activities, is an intellectual activity that has its own "rules of the game". Only if the study complies with these rules is it meaningful (in different ways as viewed by different people). At present, there are at least three effective methods of studying the history of science in accordance with the rules.

The first method is positivist chronicling. This method has been used for the study of ancient history for a long time, and prevails in the study of the history of science. It shares some attributes with the method of textual criticism of the School of Qianjia, which existed during the reign of Emperor Qianlong (1736–1795) and Emperor Jiaqing (1796–1820) in the Qing Dynasty (1636–1912). Chronicling refers to evaluating and recording historical events in a timeline in order to faithfully come up with an outline of history. The aforementioned books *The History of Chinese Astronomy* (1955) and *The History of Chinese Astronomy* (1981) are representative works of this method. In fact, this approach is adopted to varying extents by studies of history featuring other methods. The disadvantage of this method is that it may turn the study into a mere "record of achievements" without any in-depth evaluation or thinking.

The second method is concept analysis, which was put forth by the Idea School. This approach was not used for the study of the history of science until the beginning of the 20th century. It features the study of original documents primarily to learn the ideas of the authors of these documents, but not to discover their achievements, focusing on intellectual development. Well-known works in this regard include *Galileo Studies* (1939) by Alexandre Koyré and *The Origins of Modern Science* (1949) by Herbert Butterfield. Butterfield was opposed to the act that turns the study of the history of science into "recording achievements". He said,

> *The whole fabric of our history of science is lifeless and its whole shape is distorted if we seize upon this particular man...who had an idea that strikes us as modern, now upon another man...who had a hunch or an anticipation of some later theory — all as if one were making a catalogue of inventions or of maritime discovers. It has proved almost*

more useful to learn something of the misfires and the mistaken hypotheses of early scientists, to examine the particular intellectual hurdles that seemed insurmountable at given periods, and even to pursue courses of scientific development which ran into a blind alley, but which still had their effect on the progress of science in general.[4]

The method of concept analysis and the results of the studies guided by this method have little influence on the study of the history of science in China. Although China has seen a lot of books with the title of "Kexue Sixiang Shi" in recent years, the title of these works should be translated into "The History of Scientific Ideas", which is a branch of the history of science, rather than "The Intellectual History of Science", which involves the method of concept analysis.

In addition to the above two approaches, there is a third approach, the sociological approach that first emerged in the 20th century. In 1931, Boris Hessen, a former Soviet historian of science presented his paper "The Social and Economic Roots of Newton's 'Principia'" at the Second International Congress of History of Science, marking the birth of this unique Marxist sociological method of the history of science. Afterwards, this method received the active support of many left-wing historians of science. Some representative works of this kind include *The Social Function of Science* (1939) by J. D. Bernal and *Science, Technology and Society in Seventeenth-Century England* (1938) by Robert K. Merton, both pioneer works of sociology and a more important genre of the study of the history of science by using the sociological approach.

6. From intradisciplinary studies to *waishi* study

In essence, none of the abovementioned three methods is superior to any other, and none can be solely applied in study. The history of ideas approach and the sociological approach, which emerged later as approaches to the history of science, have indeed shed new light on the research in this field because they are both fare superior to the archaic chronicling method. As for the relationship between the two methods

---

[4] Herbert Butterfield. *The Origins of Modern Science*, New York: Collier Books. Translated by Zhang Liping *et al.*, Huaxia Publishing House, (1988), pp. 2–3.

and their roles, Wu Guosheng, a professor at Peking University, has given a good explanation:

> *As two ways of explaining the development of science, the history of ideas approach and the sociological approach do have something unique to offer, but also have some shortcomings. Although these defi-ciencies have been elaborated, a more sophisticated method of intradis-ciplinary studies and waishi study has not yet appeared. Perhaps it is impossible to replace the two methods with a new comprehensive one. Perhaps in studying the development of science, they are the foundation, on which a complete history of science can be established.*[5]

Let's return to the issue of intradisciplinary studies and *waishi* (*out-side history*). In fact, the traditional chronicling method used to be the one and only way of intradisciplinary studies, which is evidenced by the works on the history of science published before. (However, some mas-terpieces still fail to adopt the perspective of the history of ideas). A successful research in *waishi* cannot dispense with the sociological method.

Shifting from intradisciplinary studies to *waishi* study is not simply introducing more study objects but transforming the train of thought and perspectives. Pure intradisciplinary studies see the history of science as the history of science itself (given the previous practices in China). The study of *waishi*, on the other hand, requires that the history of science be considered as an integral part of the history of the entire human civiliza-tion. As a result of open-mindedness and different perspectives, the same study object is placed in a different background, thus presenting quite a different status and meaning.

7. Chinese astronomy and its role in the history of science

Before the 1980s, researches on the history of Chinese astronomy were characterized by two major features: making full use of the principles

---

[5]Wu Guosheng. *Kexue Sixiangshi Zhinan* (*A Guidebook on the History of Scientific Thoughts*). Chengdu: Sichuan Education Press, (1994), p. 11.

and methods of modern astronomy to ensure a modern scientific ideology in researches; carrying on the textual criticism of the Qianjia School, the national heritage, and tradition in the hope of boosting national self-esteem and self-confidence. The two features determined the topics and styles of researches. Subjects on intradisciplinary studies dominated, and source criticism, checking calculations, and interpretation of the previous achievements of Chinese astronomy studies formed the gist of these researches.

Over decades and after many years of exploration, researches on the history of Chinese astronomy have made great achievements in terms of intradisciplinary studies. Many of the research achievements are remarkable and worth noting. Xi Zezong, Bo Shuren, Chen Meidong, and Chen Jiujin are important figures in this field. Works such as the abovementioned two books with the same name of *Zhongguo Tianwenxue Shi* (*The History of Chinese Astronomy*), *History of China's Star Observation* by Pan Nai, *Collected Works of Chen Jiujin* by Chen Jiujin and *Study on Ancient Astronomical Calendar* by Chen Meidong are the epitome of the research in this field. It is worth mentioning that the *New List of Novae* authored by Xi Zongze in 1955, and its sequel, comprehensively compiled the records of novae and supernovae outbursts and their exact positions made in ancient China. They were valuable and irreplaceable historical materials for the development of astrophysics in the 1960s and became the most well-known work of China's astronomical research in the world (refer to Section 3 in Chapter 13). The books also gave birth to a new branch of the research on the history of Chinese astronomy — sorting out ancient astronomical records for modern research.[6] The great achievements made in Chinese astronomy research have made it a leading area in the research on Chinese history of science till date.

---

[6] As Chinese records of astronomical phenomena in old times were obtained across a long time, sound in classification and vast in number, they attract many Chinese and foreign scholars to research into this branch. However, it is pointed out that the work of modern astronomic projects based on these ancient materials does not belong to the research of history of astronomy, but the research of modern astronomy. Of course, it will not make trouble to consider such distinction just as a concept game and not take it seriously.

8. The trend of *waishi* study since 1990s

As the intradisciplinary studies of Chinese astronomy gradually formed a complete system, little room was left for any groundbreaking achievements. Because the basic pattern and the framework have been established, what the successors can do is only to put icing on the cake. Also, works such as *A New Catalogue of Ancient Novae* are too exceptional to surpass. Moreover, even with deepening study, many of the problems cannot be solved solely within the scope of intradisciplinary studies.

In the 1980s, a new trend in Chinese astronomical research emerged due to some changes in the overall environment at home and abroad. On one hand, after the "Cultural Revolution" (1966–1976), a new generation of students entered the field of the history of science. The changes of time did exert some influence on their professional study — they often did not like the Qianjia School, which of course did not mean that it was inferior. They were not interested in insignificant subjects, so they craved innovation. On the other hand, China's research on the history of science was prompted to embrace that of the world after the reform and opening up and thus saw a new trend emerging from the outside world. This trend is a shift to *waishi* study in the research on the history of science, namely perspective conversion. It focuses on the interaction between the development of science and society and culture. For example, the sixth International Seminar on Chinese History of Science held in 1990 in Cambridge, UK, saw three reports being presented, among which two were on the themes of "The Relationship between Ancient Chinese Astronomical Mathematics and Society and Politics" and "The Social Organization of Ancient Chinese Medicine". This was undoubtedly a sign highlighting *waishi* study.

Under the influence of the new trend, my book *Tianxue Zhenyuan* (*The Truth of the Sciences of the Heaven*) published in 1991 was well received by scholars at home and abroad, and it exceeded my expectation. I thought that some "radical" conclusions in the book might not be easily accepted soon, but the response proved me wrong. *Tianxue Zhenyuan* was reprinted in 1992, 1995, and 1997 and published with traditional Chinese characters in Taiwan in 1995. *Tianxue Zhenyuan* was rated as an important reference among works of *waishi* study (including in Master and Doctoral dissertations) published in recent years in China. Wann-Sheng Horng, an academician at the International Institute for the History of Science and professor at National Taiwan Normal University, delivered a lecture "Introduction to *Tianxue Zhenyuan* and Discussion of China's research on History of Science and its Prospect" when teaching the course "Chinese History of Science and Technology" in Tamkang University,[7] and called it pioneering research on the history of Chinese astronomy. I'm flattered by his remark. The fact that my views in the book are so widely accepted may imply that the study of outside history in China has entered a new stage.

9. Sino-foreign communication in astronomical history

With the influence of study of *waishi* (outside history), exchanges between eastern and Western astronomies in ancient times and their comparative studies have also drawn increasing attention. However, before this pleasant change, most of the conclusions concerning this aspect were made by sinologists from the west, and in Japan, while Chinese scholars rarely made important achievements (such as Guo Moruo with his *Shi Zhi Gan* (*Interpreting Heavenly Stems and Earthly Branches*), focusing on the relations between Chinese and Babylonian astronomy). Even those few research results were not usually made by professional researchers of astronomical history. This situation did not see remarkable changes until the 1980s when Chinese and foreign academic periodicals began to publish some theses including the Western astronomy highlighted by missionaries from Society of Jesus in China at the end of the Ming Dynasty and its origin; the connections between ancient Chinese astronomy and that of Babylon, India, and Egypt in ancient times; and the relationship between Chinese and Islamic astronomy in old times.

The reason why the research of astronomical history could occupy such an important place in the history of cultural exchanges in ancient times is that astronomy was the only exact science at that time. Among those ancient cultural exchanges, despite the difficulty in ascertaining whether some elements in a culture were created by its own people or inspired by

---

[7] See *The History of Science Newsletter* (Taiwan). No. 11, (1992).

others, the elements related to astronomy such as star catalogues, astronomy equipment, astronomical parameters, etc., were considered to be highly important and evolved, thus providing some obvious clues to research into the intricate cultural exchanges at that time.

Besides, studies on the history of astronomy can help history and archaeology fix the timing of events that have happened in the past. As astronomical phenomena in old times can be retraced by employing modern astronomy skills, the year in which certain historical events took place can be ascertained (such as King Wu's attack of King Zhou) using these astronomical details. It can even be used to identify during which dynasty those ancient books (such as *Zuo Zhuan*, or *The Commentary of Zuo*) recording astronomical phenomena were finished. This function of the research of astronomy history can be best demonstrated by nine special topics on astronomy history in the "*Xia, Shang and Zhou Chronology Project*", a key program for science and technology development of China during the 9th Five-Year Plan Period.

In terms of the research of history of religion, astronomy has also begun to play an increasingly important role in this regard. In the past, the dissemination of religions usually depended on astrology and astronomy to draw attention of the public and rulers. For instance, a long time back, Buddhism (especially Tantrism) was introduced into the Central Plain during the Sui and Tang dynasties (581–907); a later example is that Christianity widely spread in China at the turn from the Ming Dynasty to the Qing Dynasty (1636–1912). The fact that some recent international conferences on the history of religion have invited specialists of astronomy history is in line with just such considerations.

All such functions and influences of the study on *waishi* have helped create more opportunities for study and made this area a hotbed for further research.

10. A bridge between two cultures

Modern civilization has been developing at such a fast speed that less attention has been paid to the connections between natural sciences and social sciences. The emergence of masterminds like Aristotle, a genius and a scientist with encyclopedic knowledge, is a million to one chance.

Certainly, this is not a pleasant fact, just the cost human beings have to pay for modern civilization. This issue has concerned some far-sighted figures. Dating back to the beginning of the 20th Century, J.B.Conant, then president of Harvard University, proposed the usage of "science and academia" to narrow the gap between science and humanities, which gained much popularity. At that time, Dr. George Sarton, who was the founder of the history of scientific society, was appealing for constructing a bridge between the sciences and humanities. This bridge he appealed for refers to nothing but the history of science, and he held that "the construction of that bridge is the main cultural need of our time".[8]

Over fifty years has passed, and yet the bridge Sarton longed for has not been put in place. Instead, there has emerged a wider gap between the sciences and humanities. Western scholars have paid much more attention to this issue than Chinese scholars. For instance, in 1959 when Charles Percy Snow delivered a well-known lecture titled "The Two Cultures and the Scientific Revolution" at Cambridge University and delved into the dangerously wide gap between the sciences and humanities, it triggered heated international discussions. While some achievements have been made in China to build the connecting bridge called for by Dr. Sarton, the progress is considered an outbuilding to the mansion of natural sciences. Most people in this field cannot understand Dr. Sarton's ideal, and even those who have contributed to its achievements do not take it as their own responsibility to build wider connections between the sciences and humanities as Dr. Sarton did.

11. Three motivations to conduct research on *waishi*

To summarize, we can identify three motivations for *waishi* studies.

First, the demand for further development of science history; second, the need of researchers to expand research fields; and third, the connections between the sciences and humanities with the human civilization as a whole.

---

[8] George Sarton. *The History of Science and the New Humanism*. Cambridge: Harvard University Press, (1962), p. 58. Translated by Chen Hengliu, *et al.* Huaxia Publishing House, (1989), p. 51.

The first two reasons come from the requirements of researchers of science history, while the third one may engage humanists in *waishi* research, which is not uncommon in other countries these days.

With the increasing popularity of *waishi* study, the research on astronomical history has integrated into the field of civilization research and cultural history, thus leading to numerous interactions between science and culture. Compared with previous studies, the historical research into astronomy covers much more knowledge, thus requiring researchers to have a better knowledge structure and higher quality of studies. Put simply, current researchers of astronomy history should not only receive the formal discipline in astronomy but also be equipped with the humanistic quality of common humanists. However, the existing wide gap between the sciences and humanities has made it difficult to replenish the team of researchers of astronomical history who possess these requirements, not to mention we live in an era when academic studies are failing to attract much attention.

Luckily, the field of the history of astronomy does not require many researchers to be involved. In this sense, there can be several "freaks" meeting the above requirements who are able to stand a needy and lonely life in China, a country with a vast area and a large population. I think just a few such persons will be enough to make a difference in this field.

Chapter 2

What Kind of People in Ancient China Needed *Tianxue?*

We look at the ornamental figures of the sky, and thereby ascertain the changes of the seasons. We look at the ornamental observances of society, and understand how the processes of transformation are accomplished all under heaven.

— *Tuan, Bi, The Book of Changes*

The sage, in accordance with (the Yi), looking up, contemplates the brilliant phenomena of the heavens, and, looking down, examines the definite arrangements of the earth.

— *Xi Ci (I), The Book of Changes*

What kind of people need *tianxue* (*the sciences of the heaven*)? Ancients and modern people might give varied answers to this question. Proceeding from modern notions, modern people may envisage the ancient scenarios and intentions of ancient people in the same light, thus easily giving rise to misunderstandings and prejudices. We must first of all have a clear idea of what exactly *tianxue* was before we can figure out what kind of people needed *tianxue* in ancient China.

1. What is the use of astronomy?

If, on March 8, 1996 (the date the author happens to write this chapter, implying nothing in particular), you asked an astronomer in China: "What on earth is the use of astronomy?", it might have been seen as a provocative question, as Chinese astronomy was undergoing a transition period, suffering from the pains of not being "the main economic battlefield", thus dealing with such issues as shortage of funds and many practitioners thinking of quitting the profession. Many people were of the opinion that astronomy is simply useless, bringing in no economic benefits. That was why astronomers were a bit oversensitive at that time.

Astronomy does have some practical utilization today, such as in fields of time service, navigation, serving spationautics, etc. Nevertheless, the biggest function of astronomy is after all "intangible", and that is "to explore nature", starting from the earth and exploring the solar system, the galaxy, and the whole universe, trying to know their origins, status quos, and evolutions. As these do not bring in direct economic benefits, shortsighted people and those anxious to get instant benefits deem it useless, which is natural and understandable. But mankind needs this kind of exploration. It has long been recognized in developed countries that the non-utilitarian utilization will be of profound use.

Therefore, in modern society, who or what needs the use of astronomy to explore nature? The answer to this can only be the society or science, in general, rather than some individual or social group.

2. The role of astronomy in ancient society

Now that astronomy has such a role and status in modern society, did it have similar role and status in ancient times? We must remember the fact that modern people have never lived in ancient society. And it is often the case that modern people talk about the ancient period with a modern mindset. Even a learned person can hardly avoid this tendency. This is also the case when it comes to the major differences between ancient and modern astronomy.

For many years, domestic scholars have deemed the following line as the golden rule when discussing the origin and role of ancient astronomy.

The first is astronomy — to decide seasons, nomadic people and agricultural nations need it absolutely.[1]

Reading this line, one may find no faults. As the words of a sage are naturally looked upon as a golden rule, researchers' mindset may be unconsciously confined. As "astronomy is for agricultural peoples to · decide seasons", it is natural that the use of astronomy in ancient China was to serve agriculture. Therefore, we have here the clichéd statements of "astronomy is to serve agriculture" and "calendar is to serve agriculture", which have long become the starting point for discussions on Chinese ancient *tianxue* (*the sciences of the heaven*), ruling out the possibilities that the astronomy calendar might also serve other purposes.

Yet, there comes a more important fact that, given the above-mentioned starting point, people would naturally see *tianxue* in ancient China as a technology that serves both production and exploration of nature (which is exactly the mission of modern technology). This, on the surface, appears well-reasoned; but as a matter of fact, it is far from historical accuracy.

3. The fallacy behind the concept that astronomy was to serve agriculture

To till land, peasants need to know about climate and other natural phenomena of seasons, which is deemed as overwhelming evidence that astronomy was there to serve farming. But those holding this view obviously ignore such a simple fact that both written records and archeological proofs have indicated that the history of agricultural history is much longer than that of astronomy. In other words, farming existed long before astronomy. Even when astronomy came into being, it did not achieve the monumental development that agriculture did.

In fact, contemporary knowledge tells us that agriculture relies little on astronomy. Peasants do not need to have full knowledge about climate and natural phenomena of seasons. A divergence of one or two days does not make any difference. The Chinese calendar has evolved over a period of 3,000 years, and this is widely recognized as the most scientific part of the ancient *tianxue* of China. It has numerous observations, calculations,

---

[1] Engels: *The Dialectics of Nature.*

formulas, and skills that are precise to the very minute, and even seconds. This is definitely not for guiding peasants to till land.[2]

In ancient times, common folks, including farmers, did not need to know much about astronomy, which is also the case in today's society, even with the increasing popularity of science. Even though farming needed to be conducted based on climate and the natural phenomena of seasons, farmers could manage to get by observing phenological changes occurring around them. In fact, ancient people were able to define solar terms through observing animals, plants, and climate phenomena for a long period of time. Among the twenty-four names given to solar terms today, twenty of them are related to seasons, climate, and phenology, which is undoubtedly a vivid example of the importance of astronomy. Later, Chinese calendars and other such records with notes on the solar terms over the period of a year emerged, making it much easier for people to know the subsequent climatic changes and natural phenomena. From the perspective of astronomy, solar terms are made based on the apparent annual motion of the sun, which is caused by the earth's movement around the sun, and solar terms seem to be systematically connected with astronomy, but although some links do exist between them, they are two radically different things. The apparent annual motion of the sun is rather complicated and abstract, and even to date, only a few scholars that are familiar with astronomy can understand it completely. Therefore, it is obviously not reasonable to assert that a man who farms according to phenological phenomena must know astronomy, just like we cannot judge that modern people who see the calendar and know solar terms are certainly familiar with astronomy.

Farmers and other common people could not understand astronomy, and the same is true of people living in modern society, where knowledge of the sciences has been widely disseminated. In this regard, a dictum by Gu Yanwu (1613–1682) in his *Ri Zhi Lu* (*Notes on the Daily Accumulation of Knowledge*), has greatly misled numerous people in today's society:

Dating back to dynasties of Xia, Shang and Zhou, everyone was familiar with astronomy. When Antares goes to the west, it will be

---

[2] For the details of the argument, please see Jiang Xiaoyuan. *Tianxue Zhenyuan* (*The Truth of the Sciences of the Heaven*), Liaoning Education Press, (1991, 1995), pp. 39, 140–145.

cooler soon, as was often said by farmers; women normally said that delta, epsilon and zeta stars of Orion can be seen through the open window indicating that a wedding is around the corner; it could be heard that as the moon approaches Aldebaran, it will rain a lot from frontier soldiers; and children would say that lambda stays between the sun and the moon when the latter two stars are going to meet each other.

Gu's first three lines are quoted from the three songs of *Shi Jing* (*The Book of Songs*), namely, *Bin Feng-Qi Yue* (*Life of Peasants of Songs Collected from Bin*), *Tang Feng-Chou Mou* (*A Wedding Song Collected from Tang*), and *Xiao Ya-Chan Chan Zhi Shi* (*Eastern Expedition of Book of Odes*), and the fourth line is from *Guo Yu-Jin* (*Discourses of the States-the State of Jin*). The three songs are written in the tone of a peasant, a wife, and a garrison soldier. But obviously it does not mean that the poems were authored by a peasant, a wife, and a garrison solider. Using the first-person narrative to create literary works is a very common concept, both at home and abroad. The identity of "I" and the profession are not necessarily the same as those of the author.

More importantly, even if the authors of the three poems were a farmer, a wife and a garrison solider, respectively (which, I'm afraid, would be mocked by scholars of classical literature), we would have no way of drawing the conclusion that "astronomical knowledge had been widespread to farmers, women and army men". Poems are supposed to express a poet's true emotions and their imagination. When they chanted about celestial phenomena, it did not mean they understood the laws of these phenomena, nor did they understand astronomy, which is rather abstract and sophisticated — far beyond the mental capacity of farmers, women, and garrison soldiers. Let us make similar inferences. A poet chanting about wind and clouds surely does not imply that he or she knows about meteorology; a poet chanting about rivers surely does not imply that he or she knows about hydraulics; a poet writing about a bronze mirror does not necessarily imply that he or she understands metallurgy and optics. The answers are not difficult to derive when you start with common sense.

Strained interpretations and farfetched analogies are misleading and not helpful for us to have a clear picture of history.

4. The implications of *tianwen* ("astronomy")

However, Gu should not bear any responsibility for his words that may mislead modern people, because it is they who misunderstand his thoughts, interpreting ancient people's ideas based on concepts from modern times. In fact, *"tianwen"* (Note: The pronunciation of *"tianwen"* in Chinese is the same as "astronomy".) from Gu's "dating back to dynasties of Xia, Shang and Zhou, everyone was familiar with *tianwen*" is a conventional name used by ancient Chinese, whose understanding is totally different from that in modern China. The essay in one of China's ancient books where the word *"tianwen"* was used for the first time is called *Ben, Tuan, the Book of Changes*:

> *"We look at the ornamental figures of the sky,[1] and thereby ascertain the changes of the seasons. We look at the ornamental observances of society, and understand how the processes of transformation are accomplished all under heaven".*

Also in *Xi Ci (I), the Book of Changes*:

> *"(The sage), in accordance with (the Yi), looking up, contemplates the brilliant phenomena of the heavens,[2] and, looking down, examines the definite arrangements of the earth".[3]*

Tianwen (astronomy) was usually mentioned using the comparison with *dili* (geography) by ancient people. It refers to pictures of celestial bodies moving in the sky, and such pictures were named as *"wen"* (For instance, Volume IX-1 in *Shuowen Jiezi* (*The Analytical Dictionary of Chinese Characters*) said that *"wen"* refers to intricate tattoo patterns, while *li* in *dili* likewise means pictures of distribution of all things on the ground (Even now, *"wenli"* is still in use, and it means patterns or lines on the surface of something). Therefore, it can be seen that *tianwen* in ancient times was actually synonymous with phenomena in the sky, and not astronomy. The dictum of Gu only indicated that ancient people were familiar with some (names of) celestial phenomena.

To gain more insights into ancient people's usage of *tianwen*, more examples can be taken from later ancient books. For example, in *Han*

Shu-Wang Mang Zhuan (*Biography of Wang Mang, the Book of Han*) it is said that:

> *"In the eleventh month (by lunar calendar), a comet appeared in (the constellation) Chang. It traveled southeastwards for five days and disappeared. (Wang) Mang several times summoned and questioned his Chief Grand Astrologer, Tsung Hsüan, and various diviners. They all answered falsely, saying, 'The astrological phenomena are peaceful and good, so that the many bandits will soon be destroyed.' Thereupon (Wang) Mang (felt) a little more tranquil".*

According to ancient Chinese astrology, that the comet can be seen near the Chang usually suggests dangers and inauspicious events around the corner (details will be provided later), but his astrologer and diviners did not tell him the truth and instead gave a false answer to relieve his anxieties. Another instance comes from the *Shu Ji* cited in *Jin Shu-Tianwen Zhi* (*Annals of Shu* cited in *Astronomical Treatise, the Book of Jin*):

> "(Wei) Ming Emperor asked Huang Quan: 'Now the whole land is in a tripartite division between Shu, Wu and us. Which is the legitimate regime authorized by the Heaven?' Huang replied: '*Tianwen* can prove it. Last time when the Mars approached Antares, (Wei) Wen Emperor passed away, but the emperors of Wu and Shu were still alive. This is the evidence.'"

That the fire star stays besides Antares was considered an astronomical phenomenon suggesting inauspicious events on the horizon in ancient China. Therefore, when this phenomenon happened and Wen Emperor of Wei died, it indicated that the emperor went with the changes of phenomena in Heaven, and accordingly it was a legitimate regime. However, nothing bad happened to Wu and Shu, implying that they were not the authorized forces.

Based on the denotation of *tianwen* — astronomical phenomena — another meaning came into existence, which was the knowledge to forecast events in the human world by observing astronomical changes. Sentences mentioned several times in *Xi Ci (I)*, *the Book of Changes*, have implied this meaning of *tianwen*. One is that, in the heavens there are (different) figures there, and on the earth there are (different) bodies formed. (Corresponding to them) were the changes and transformations exhibited (in the Yi), and another is that (the sage), in accordance with (the Yi), looking up, contemplates the brilliant phenomena of the heavens, and, looking down, examines the definite arrangements of the earth — thus, he knows the causes of darkness (or, what is obscure) and light (or, what is bright). Apart from this, there is another discourse in this book providing the exact explanation of this meaning of *tianwen*, and this has, as usual, been ignored by previous works of the history of science:

> *"Therefore Heaven produced the spirit-like things, and the sages took advantage of them. (The operations of) heaven and earth are marked by (so many) changes and transformations; and the sages imitated them (by means of the Yi). Heaven hangs out its (brilliant) figures from which are seen good fortune and bad, and the sages made their emblematic interpretations accordingly. The He gave forth the map, and the Lo the writing, of (both of) which the sages took advantage".*

Here, both the map (*Hetu*) and the writing (*Luoshu*) are sacred objects by nature, and *"Heaven hangs out its (brilliant) figures from which are seen good fortune and bad"* refers to changes between the sky and the ground; therefore, if sages, namely rulers, can fully take advantage of them, they can understand how to govern their countries. Simply speaking, rules to handle human beings and manage the country can be created and understood by imitating the laws of changes in nature. These statements, though mysterious and incredible to modern society, were political ideals of ancient people who firmly believed in them. With regard to *tianwen*, the postscript of *"21 Schools of Tianwen* "of Ban Gu's *Han Shu-Yi Wen Zhi (Treatise on Literature, the Book of Han)* says:

> *"Tianwen is an activity: it forecasts the happening of blessed or inauspicious events by observing the changes of the twenty-eight constellations and the five planets (Venus, Jupiter, Mercury, Mars and Saturn), which offered emperors counsels on ruling".*

In fact, Ban Gu discussed the nature of various kinds of knowledge in his *Treatise on Literature in the Book of Han,* which simply represented the general attitude of the cultural field in ancient China. The nature of *tianwen* that he defined in his book was the same as the traditional opinions of it in the following two thousand years.

From the above discussion, it can be easily understood that in ancient China, the second meaning of *tianwen* was actually a synonym for "astrology" used in modern times, instead of meaning today's "astronomy".

Careful readers may have noticed that the word used in this chapter is not astronomy but *tianxue*. During the last five years when I embarked on writing *Tianxue Zhenyuan*, I started to use *tianxue* a lot when talking about relevant facts of ancient China in some books and essays. Of course, it was not because I took delight in being different, but I did not want to confuse these two concepts. Afterwards, some other scholars from this field also began to use this word.

5. Was there any astronomy in the modern sense in ancient China?

"Astronomy" today falls under the concept of modern science, a discipline of modern times. We must conduct in-depth studies rather than rely on assumptions if we are to find out whether the discipline existed in ancient China. Likewise, we cannot draw the conclusion that modern chemistry was no stranger to people in ancient times, though the discipline can date back to alchemy, an ancient practice.

However, no direct account has been given on the subject since modern astronomy is generally believed to be a discipline of the past. It, therefore, makes no sense to argue about its very existence. As a matter of fact, many discussions have been made about astronomy's role in ancient times, yet most of the remarks are extremely unpleasant to the ear. For example, Mathew Ricci, a missionary of the Society of Jesus who visited China in the late 1500s, said,

> *"They are concentrating on astronomy, or what is known as astrology in the science community. They believe all that is happening in China depends on horoscope".*[3]

---

[3] Mathew Ricci. *Mathew Ricci China Sketches* [M]. Translated by He Gaoqi, *et al.* Zhonghua Book Company, (1983) p. 22.

M. Delambre, a French Scholar, said, *"Astronomy has never happened in China, though the country has a long history"*.[4]

A. Sedillot's words sound even more unpleasant. He said:

> *They have never been freed from being slaves of superstition or astrology. Chinese people don't need their curiosity to observe the star-studded sky so as to gain complete understanding of its rules and causes. Instead, they devote their impressive stamina to worthless talks about astrology. This is a tragic consequence of a savage custom".*[5]

Remarks of cultural superiority and aggressiveness are bound to repulse Chinese academia. And rightly so. Despite some elements of superstition in astrology, it still played a positive role in ancient society. This discipline was one of the very few exact sciences that was in existence at that time.[6] We cannot equate astrology to astronomy, but the former cannot be present without knowledge of the latter. This can be evidenced by the fact that one needs to calculate the exact position of the sun, the moon, and five great planets with the laws of astrology at a given time.[7] It makes sense to draw parallels between astrology and astronomy in a sense given that astrology has both leveraged on and given a strong boost to astronomy.

That is why we use "Ancient Chinese *Tianxue*" in a bid to avoid confusing these concepts and to serve as a reminder of how ancient astrology and astronomy are connected to each other.

6. *Zhou Bi Suan Jing (The Arithmetical Classic of the Gnomon and the Circular Paths of Heaven)* — The earliest attempt of axiomatization

Does *tianxue* (*the sciences of the heaven*) give top priority to exploration of the nature rather than that of earthly gains since the discipline in

---

[4] Quoted from Zheng Wenguang. *Origin of China's Astronomy* [M]. Science Press, (1979) pp. 6–7.

[5] *Ibid.*

[6] Refer to Jiang Xiaoyuan. *Astrology in History* [M], Shanghai Science and Technology Education Press, (1995), pp. 271–274.

[7] For detailed argumentation, please refer to: *Tianxue Zhenyuan* [M], pp. 151–167.

ancient China was barely related to agriculture? No it does not. It is by no means easy to figure it out. Many people in modern society tend to take things for granted, trying to imagine ancient people's thinking from a modern perspective. They believe that *tianxue* is supposed to probe into the nature as its main task because that is what modern astronomy does. Built on a modern mindset that values "science" rather than "superstition", people tend to beautify their ancestors by imposing their opinions on them.

In most ancient civilizations, including Egypt, Babylon, India, and Maya, among others, astronomical knowledge has evolved along with astrology. Astrology is supposed to serve political, social, and spiritual life. While astrology is not used for "economic benefits", it is in principle a far cry from modern science, which works to tap into nature.

The only exception was ancient Greece. Scholars believed that nature-exploration-oriented and independent astronomy had already come into being and flourished well before astronomy, which could be traced back to ancient Babylon, was introduced to Greece. Astrology was brought to Greece by Berossus in about 280 B.C.[8] This is a profound exception because it is safe to say that ancient Greece is the spiritual source of modern astronomy and the wider scientific system. As such, we stand to benefit from reviewing the "words of sage" from *Dialectics of Nature* by Engels (1820–1895).

If modern science wants to look back into how its general principles have evolved, it has to review what was happening in Greece.

Proponents and opponents are enormously divided over the history of science in their minds.

Will ancient Chinese *tianxue* be a second exception? No, it will not be, at least not now. Attempts at axiomatization and geometric universe modeling from *Zhou Bi Suan Jing* (*The Arithmetical Classic of the Gnomon and the Circular Paths of Heaven*) were short-lived ones.[9] Apart from the above exception, *tianxue*, among other kinds of knowledge, was highly practical. Yet *tianxue* distinguished itself because it assumed highly sacred missions in ancient Chinese society.

---

[8] Refer to Jiang Xiaoyuan: *Astrology in History*, pp. 59–60.

7. Chinese emperors and the need for astronomy

Now we are approaching the subject: What was the divine mission of *tianxue* in ancient China? Or what kind of people in ancient China needed *tianxue*?

Actually some clues have been revealed in the previous sections. First, we need to find out who needed to make astronomical observation and geographical surveys. The *Xi Ci (II) of the Book of Changes* clearly tells us who they are. They are listed in the following.

Bao Qi (Fu Xi or Fu Hsi), heavenly king in ancient China, who, looking up, contemplated the brilliant phenomena of the heavens, and, looking down, examined the definite arrangements of the earth.

The Xi Ci (II) gives a simple and idealized picture of Confucianism and about the evolution of ancient civilization, which ranks Fu Xi first among all the five kings and sovereigns, which included Fu Xi, Shen Nong, Huang Di (or Ji Xuanyuan), Yao, and Shun respectively.

These *deuses* and sovereigns are seen as creators of things and ideas in a civilized society. We need to understand who stands out (brilliant) among these figures, from which are seen good fortune and bad. *Xi Ci (I)* says it was a sage, the ruler. In *Tianguan Shu of Shih Chi (Book of the Tianxue Officials, The Records of the Grand Historian)*, Sima Qian elaborates more clearly the interaction between heaven and mankind and the *tianxue* needed by sages.

The Grand Historian said:

> *Since the creation of mankind, emperors in the world had all tried to make a calendar according to the movement of the sun, the moon and the stars. It was not until the Five Emperors and the Three Dynasties (Xia, Shang and Zhou) that calendar making had been well inherited and carried forward. The central area was an area with well-developed cultures while the external regions were where barbarians resided. They divided the territory into 12 divisions, contemplated the brilliant phenomena of the heavens, and examined the definite arrangements of the earth. Then they knew there are the sun and the moon in the sky and yin and yang on the earth, five planets in the sky and five elements on earth, constellations in the sky and prefectures and states on earth, which are all corresponding to each other. The three lights of the sun, the moon and the stars are incarnated by the energy of yin and yang on earth. As the three lights are*

all originated from the earth, sages unify the heaven and the earth and govern them together.

In the early phase of ancient Chinese civilization, *tianxue* played an extremely important role in politics. It was a matter of primary importance, and even the only matter, that counted for emperors. This can be confirmed by the earliest historical records in China.

Shangshu (*The Book of Documents*) is a classic work of Confucianism. In the eyes of modern people, it is a collection of political documents (including transcripts and adaptations). *Yao Dian* (*The Canon of Yao*) is the first article of *Shangshu*, which tells about why to write *Shangshu*. The *Shuxu* (*The Preface of the Book of Documents*) reads:

> *Emperor Yao was good at knowing outstanding people and using them to the full of their potentials. His virtues were so lustrous as to be known to the world people. When he was about to retire, he decided to hand over the crown to Shun (also known as Yu Shun). Hence, the Canon of Yao was written to record this feat.*[10]

The Canon of Yao records the essentials of governance of Emperor Yao when he was on the throne, as well as Yao's directions and decrees in cultivating Shun as his successor. Half of the some-400-character article tells of Yao's achievements when he was on the throne.

Examining into antiquity, (we find that) the Di Yao was styled Fangxun. He was reverential, intelligent, accomplished, and thoughtful — naturally and without effort. He was sincerely courteous, and capable of (all) complaisance. The bright (influence of these qualities) was felt through the four quarters (of the land), and reached to (heaven) above and (earth) beneath. He made the able and virtuous distinguished, and thence proceeded to the love of (all in) the nine classes of his kindred, who (thus) became harmonious. He (also) regulated and polished the people

---

[10] Opinions vary regarding who wrote the *Shuxu* (*The Preface of the Book of Documents*). Some say it could be Confucius, a scholar of the Qin State of the Zhou Dynasty, Shi or a scholar of the Han Dynasty. See also Jiang Shanguo: *An Overview of Shangshu*, Shanghai Chinese Classics Publishing House, (1988), pp. 63–65. It is obviously not important discussing who wrote it here.

(of his domain), who all became brightly intelligent. (Finally), he united and harmonized the myriad states; and so the black-haired people were transformed. The result was (universal) concord.

He commanded the Xis and Hes, in reverent accordance with (their observation of) the wide heavens, to calculate and delineate (the movements and appearances of) the sun, the moon, the stars, and the zodiacal spaces, and so to deliver respectfully the seasons to be observed by the people. He separately commanded the second brother Xi to reside at Yu-yi, in what was called the Bright Valley, and (there) respectfully to receive as a guest the rising sun, and to adjust and arrange the labours of the spring. "The day", (said he), "is of the medium length, and the star is in Niao — you may thus exactly determine mid-spring. The people are dispersed (in the fields), and birds and beasts breed and copulate".

He further commanded the third brother Xi to reside at Nan-jiao, (in what was called the Brilliant Capital) to adjust and arrange the transformations of the summer, and respectfully to observe the exact limit (of the shadow). 'The day,' (said he), "is at its longest, and the star is in Huo — you may thus exactly determine mid-summer. The people are more dispersed; and birds and beasts have their feathers and hair thin, and change their coats".

He separately commanded the second brother He to reside at the west, in what was called the Dark Valley, and (there) respectfully to convoy the setting sun, and to adjust and arrange the completing labours of the autumn. "The night" (said he), "is of the medium length, and the star is in Xu — you may thus exactly determine mid-autumn. The people feel at ease, and birds and beasts have their coats in good condition".

He further commanded the third brother He to reside in the northern region, in what was called the Sombre Capital, and (there) to adjust and examine the changes of the winter. "The day", (said he), "is at its shortest, and the star is in Mao — you may thus exactly determine mid-winter. The people, keep in their houses, and the coats of birds and beasts are downy and thick".

The Di said, "Ah! you, Xis and Hes, a round year consists of three hundred, sixty, and six days. Do you, by means of the intercalary month, fix the four seasons, and complete (the period of) the year. (Thereafter), the various officers being regulated, in accordance with this, all the works (of the year) will be fully performed".

It is worthwhile to ruminate over the ancient records, as quoted above. The first paragraph is a sheer eulogy of Yao's merits and virtues in managing family relations, people-to-people relations, and state-to-state relations. This coincides with the famous saying of a much later time *"First free yourself of wrong-doings and evil thoughts, then bring order to your family, after which govern your people well and the land is yours"*. The second paragraph is a record of only one merit of Emperor Yao, who has been idealized and seen as a sage for thousands of years, and this merit is that he appointed four astronomers to the four directions of the land to observe the celestial phenomena and determine how to make a calendar.

Though *The Canon of Yao* was made to mark Yao's abdication of the throne to Shun, the imperator attached such great importance to *tianxue* affairs that he only recorded this in *The Canon of Yao* and did not make any mention of other affairs, be they political, military, or diplomatic, which modern people find hard to understand.

Some people may wonder whether *The Canon of Yao* is incomplete, and this is only a speculation not worth noting. For 2,000 years, scholars and officials of Confucian classics have never put forward a similar speculation. The *Wudibenji of Shih Chi (Basic Annals of Five Emperors, Records of Great Historian)* also recorded only two stories about Emperor Yao, namely, arranging *tianxue* affairs and abdicating the throne to Shun. In other words, although Emperor Yao accomplished numerous feats, only these two things have been eulogized by later generations. Therefore, we can well envisage the importance of *tianxue* in ancient politics in China.

If you hold that the stories about Emperors are not credible as they are based on isolated evidences, then we can take a look at what Shun did when he first took the throne. The *Wudibenji of Shih Chi (Basic Annals of Five Emperors, Records of Great Historian)* recorded stories about Emperor Shun's regency.

By then Emperor was becoming senile. He ordered Shun to act as regent so as to observe whether he suited the divine will. Shun then observed the Big Dipper to see whether there was anything abnormal with the movement of the sun, the moon, Venus, Jupiter, Mercury, Mars and Saturn.

Despite the controversies in understanding the latter part of this script for 2000 years, there is no doubt that it is about the arrangement of astronomical affairs. In *Shih Chi*, there is more than one record of Emperor Shun's feats of governance, but the first matter recorded is related to *tianxue*, providing ample proof that *tianxue* was of overriding importance for ancient emperors.

8. The one who knows the heaven is the king: Essentials of ancient chinese political astronomy

The ancient emperors' need for *tianxue* was surely not because they loved science or they wanted to help farmers with their farming. So, what did they want *tianxue* for? This is a very important question that, however, has long been neglected by experts of astronomical history and historians. This is the major issue that I was trying to address in my *Tianxue Zhenyuan*. Here I will refrain from giving another detailed argument. I will just give the outline, though some arguments may not sound so convincing in this case.

In ancient China (and other civilizations), the establishment of sovereignty required more than sufficient military and economic power. *Tongtian,* which means possessing the power of communicating with the divine, is of extreme importance and absolute necessity. Without any knowledge of modern materialism, the ancient people adamantly held that they could and had to communicate with the willed and sentient heaven; hence the notion "people with the special capability is hailed as the king". Dong Zhongshu of the Han Dynasty said in *Wangdao Tongsan (King Way) of Chun Qiu Fan Lu (Luxuriant Dew of the Spring and Autumn Annals):*

The inventor of the character king (王, wang) joins three horizontal strokes, which represent the heaven, the earth and the people respectively, with one vertical stroke, and there makes wang, the "king" in Chinese character. Who else but the king could understand the connection between the heaven, earth and people under the sun?

Ban Gu's saying that "his highness" employed tianxue to "help run the state" meant just the same. Other scholars like Yang Xiangkui (1910–2000) and Kwang-chih Chang (1931–2001) had referred to the same point. Based on his archeological findings of the Xia, Shang, and

Zhou dynasties, and studies on bronze sacrificial vessels and the patterns on them, Kwang-chih Chang concluded that these bronze sacrificial vessels were all devices to communicate with the heaven, without which the authority of the reigning monarch could not be recognized.

However, among all communication means tianxue (the sciences of the heaven) is one of the most essential and direct one, and includes sets of practices like Lingtai (setting up observatory terrace), Yixiang (observing astronomical phenomena), Zhanxing (divining by astrology), Wangqi (observing air currents), and Banli (issuing calendar). It was all of these that enabled the king to declare his capability of talking with the heaven and be able to fulfill the divine mandate of his ruling the country. Therein lies the reason why Yao and Shun took up tianxue and made these practices pressing matter.

We all know that one nation cannot have two queens. The interwoven relationship between tianxue and the kingship gave it a special distinction in ancient China — that it must be tightly gripped by the imperial family, who obviously enjoyed exclusive kingship. Therefore, the rebels made their own tianxue practices when trying to overthrow the established authority regardless of the law. The story of Ji Chang (1152 B.C.–1056 B.C.) building the spirit terrace in the Lingtai (The Wondrous Park) in Daya (Book of Epics) in Shi Jing (The Book of Songs) is the instance of a feudal princess covertly practicing tianxue. However, after taking the crown, the new emperor must give orders forbidding others from touching tianxue practices in his turn. The monarchs of many dynasties often repeatedly stated the prohibition of tianxue practices in civil society — even tianxue-related books or instruments could lead to harsh sentences or death penalty, and at that time informers were encouraged with hundreds of thousands of copper coins. To put it simply, tianxue in ancient China was a pressing matter for power seekers but a forbidden domain for the established authority.

Although this was a pervasive situation in earlier dynasties, it still did not see a substantial change by the start of the Ming Dynasty. As the civilization evolved, demands on the material side grew when establishing reigning authority and tianxue, the used-to-be prerequisite of kingship, was reduced to an emblem and then further downgraded to an ornament. But Chinese people stick to their traditional values. Since their forefathers put tianxue on an important and divine footing, they have managed to maintain these values for thousands of years. Although the prohibition of practicing tianxue in civil society started to loosen since the end of the Ming Dynasty (1368–1644), the hallowed status of tianxue did not start to decline until the end of the Qing Dynasty (1636–1912).[11]

---

[11] Please refer to the third chapter *Tianxue and the Kingship* of *Tianxue Zhenyuan* about the detailed foundations and analysis of the conclusions in the brief summary.

Chapter 3

What Kind of People Were Engaged in *Tianxue* in Ancient China?

1. The tradition of state-run *tianxue* study

Whether there was traditional private or individual astrology research in the ancient civilizations could be a matter of great importance, because it might be inherently linked to modern astrology. In history, this tradition existed in ancient Greece, and the successor of its spiritual wealth — Europe (though sometimes the tradition did not show itself clearly). Take a look at the astronomers in ancient Greece and you will find that people like Thales in Ionia, Aristarchus in Samos, Eudoxus in Cnidos, and the subsequent Hipparchus and Ptolemy were not official astronomers. According to legend, Eudoxus even had his own observatory.

However, the tradition is not easy to identify in ancient oriental civilizations. In feudal autocracies, *tianxue*-related affairs were state operated, and the most typical case was that in China, which has a 3000-year history of state-run *tianxue*. And the starting point of *tianxue* in China could date back to an earlier time if we believe the historical records in *Shangshu — the Canon of Yao* (now there seems no reason to doubt it) where Emperor Yao ordered official astronomers to observe astro-phenomena in the North, the South, the East, and the West. Nevertheless, unlike ancient Greece, private *tianxue* activities, as mentioned in Chapter 2, had not been permitted until the middle period of the Ming Dynasty (1368–1644). The only exception emerged in the Northern and Southern Dynasties (402–581), and more details are provided in Chapter 4.

2. Wu Hsi (or shaman) — The earliest *tianxue* expert

Considering *tianxue*'s exclusiveness to the imperial family, it is not difficult to imagine that those engaged in *tianxue* activities were not ordinary people, and this is generally true. But the difference is that it evolved over a long period. Before the Xia Dynasty (around 2070 B.C.–1600 B.C.), people working on *tianxue* were regarded sacred. (For example, King Shun used the jade-decorated astro-observation instrument in person to know about the laws of movements of planets.) But thousands of years later, in the Ming Dynasty (1368–1644) and the Qing Dynasty (1636–1912), those *tianxue* officials were lowered to the status of mediocrity. In fact, the development of the role of these people corresponds to that of *tianxue*, which started as a prerequisite to the establishment of sovereignty, then became a symbol of the throne, and finally the decoration of the throne. Important things are done by elites and unimportant ones by common people — this is a law applicable at all times and in all countries.

Who were the first group of people working on *tianxue*? They were not scientists in the eyes of modern people. They were called Wu Hsi (shamans) — the name given to wizards and witches combined — which is exactly on the opposite side of the spectrum from scientists.

With the function of communicating among heaven, earth, people, and gods, Wu Hsi (shaman) as an occupation is, in modern people's eyes, superstitious and deceptive. However, in the early Xia Dynasty (about 4,000 years ago), shamans were viewed as sacred enough because of this very ability and were even regarded as quasi-gods. At that time, the profession of shaman was closely related to the power of king. According to research done by Kwang-chih Chang and other scholars, during the early Xia Dynasty, an emperor tended to be the leading shaman of the nation. Kwang-chih Chang said:

All the actions by founders of the Xia Dynasty, the Shang Dynasty (1600 B.C.–1046 B.C.) and the Zhou Dynasty (1046 B.C.–221 B.C.) were added some witchcraft and supernatural touches. For example, Xia Yu

*(founder of the Xia Dynasty) had the super power of preventing floods
and his "Yu Bu" (the walking manners Xia Yu showed when he performed witchcraft) became the particular walking manners exclusive to
shamans. Another example is Tang's (King Tang of the Shang Dynasty)
ability to pray for rain. According to the records in Bu Ci (oracle bone
inscription), King Tang of the Shang was the only person with the right
to prophesy during that times. In addition, Bu Ci described Shang's
dancing for praying rain and interpreting dreams. All the activities were
made by the King and the shamans. They show that the kings of the
Shang were shamans.*[1]

We have discussed the political conception that a king is the one capable of communicating with the heavens. Hence it is natural for a wizard to
become a king. Emperor Yao's ordering officials to observe astronomical
phenomena, King Shun's watching of the astronomical instrument, and
the kings' ability of praying for rain during the Shang Dynasty are all of
the same nature. Emperors and kings in later ages tended to preside over
different rituals, and this was, more or less, a practice inherited from
ancient times.

An emperor had to deal with many worldly affairs, which made it
impossible for them to fully perform their role as shamans. Therefore, the
occupation of Wu Hsi (shaman) emerged so that the emperor could fulfill
his duties. With the role of Wu Hsi, the emperor now had the ability to
communicate with heaven, and their kingship was established this way.

The earliest historical clues about the mysterious shamans can be
found in *Tianguan Shu of Shih Chi (Book of the Officials in Heaven, The
Records of the Grand Historian)*, and this can be attributed to the author
Sima Qian. In this book, he presented a list of *tianshu* (predestination)
experts in history"

Pre-Gaoxin Period (before 2480 B.C.): Chong, Li
Tang-Yu Period (Yao and Shun Period) (around 3076 B.C.–2029
B.C.): Xi, He
Xia Dynasty (around 2070 B.C.–1600 B.C.): Kun Wu

---

[1] Kwang-chih Chang: *Art, Myth and Ritual: The Path to Political Authority in Ancient
China*. Liaoning Education Press, (1988), p. 33.

Shang Dynasty (1600 B.C.–1046 B.C.): Wuxian
Zhou Dynasty (1046 B.C.–221 B.C.): Shi Yi, Chang Hong
Song State (1040 B.C.–286 B.C.): Zi Wei
Zheng State (806 B.C.–375 B.C.): Bi Zao
Qi State (11th century B.C.–221 B.C.): Gan Gong
Chu State (1042 B.C.–223 B.C.): Tang Mei
Zhao State (403 B.C.–222 B.C.): Yin Gao
Wei State (403 B.C.–225 B.C.): Shi Shen

My exhaustive research process on this list is presented in *Tianxue Zhenyuan*, so I will refrain from repeating any details here, but I will briefly share the research results and significance in what follows.

The people on the list can be divided into two categories, with Wuxian (a shaman) in the middle. Those figures in the first category, namely, Chong, Li, Xi, He, and Kun Wu, were all the shamans during the early period of the Xia Dynasty whose stories are mostly legendary, while these of the latter part (starting from Shi Yi) were all the astrologists during the Spring and Autumn and Warring States Period (770 B.C.–221 B.C.) whose stories are precisely recorded. Though Wuxian was a famous shaman during the Shang Dynasty, he was later viewed as an incarnation of shamans, thus sharing the same name with shamans.

The significance of Sima Qian's list of *tianshu* experts lies in its revealing of the relationship between ancient astrology and astronomy and shamans. The origin, in other words, the evolution of the identity of these *tianxue* experts is as follows:

shaman → astrologist → astronomer

Likewise, some of the subsequent historians also provided lists similar to Sima Qian's, such as the one from *Astronomical Treatise, the Book of Jin (1)*:

Gaoyang Period (specific time period unknown): Chong, Li
Tang-Yu Period: Xi, He
Xia Dynasty: Kun Wu
Shang Dynasty: Wuxian
Zhou Dynasty: Shi Yi

Lu State(1122 B.C.–256 B.C.): Zi Shen
Jin State (11th century B.C.–349 B.C.): Buyan Period
Zheng State: Bi Zao
Song State: Zi Wei
Qi State: Gan De
Chu State: Tang Mei
Zhao State: Yin Gao
Wei State: Shi Shenfu

This version is almost the same as that proposed by Sima Qian. Some astrologists, like Zi Shen and Bu Yan, were added, and these are all real people with recordes documented in historical texts like *Zuo Zhuan* (*The Commentary of Zuo*) and *Guo Yu* (*Discourses of the States*).

3. Full-time experts and part-time astronomers of the imperial court

The position of *tianshu* experts that originated from shamans and the officials to Grand Astrology Administrations throughout history were full-time posts. But there were some special cases where chief astronomers held some other posts simultaneously, and this phenomenon was seen from early times. Shi Yi and Chang Hong are two such people, whose names were in the list provided by Sima Qian. Shi Yi was the astrologist Yin Yi during the reign of Emperor Wen (1122 B.C.–1125 B.C.) and Emperor Wu (1087 B.C.–1043 B.C.) in the Zhou Dynasty.[2] Though astrologists in ancient times enjoyed a high status, almost equivalent to that of the emperor, they had to take charge of more affairs besides *tianxue,* a setting that was quite different from the later imperial astronomers who were responsible for *tianxue*-related affairs alone.[3] Chang Hong was the one in charge of *tianxue* affairs for the imperial court during the Zhou Dynasty,[4] despite being just an ordinary senior official.[5] This phenome-

---

[2] Jiang Xiaoyuan: *Tianxue Zhenyuan* (*The Truth of the Sciences of the Heaven*), (2007) pp. 69–98.
[3] Guo Yu (*Discourses of the States*) commented by Wei Zhao.
[4] More details are provided in Chapter 4.
[5] In *Shih Chi Jijie*, the author, Pei family, quotes a word by Zheng Xuan: "Changzhong, senior official of the royal family in the Zhou Dynasty.

non became more prevalent since the Han Dynasty. Take Cui Hao in the Northern Wei Dynasty (386–534) as an example. Cui was publicly regarded as the sole authority of *tianxue,* and his explanations of great astronomical phenomena were well received,[6] despite the position of imperial astronomer being held by someone else. Another example is Xu Guangqi (1562–1633) of the late Ming Dynasty. He was undoubtedly the most important and the most active figure in *tianxue* in the royal court, but he was not the director of the Grand Astronomy Administration.

Although there are many more cases of this kind, the phenomenon was not that frequently seen in history. One of the fundamentals contributing to *tianxue*'s historical status is in the seemingly official titles it brings (like the imperial *tianxue* academy and its director) and its actual content. Also, there were no strict standard in choosing *tianxue* officials. More information about this can be discovered from a list of ancient astronomers provided by Li Chunfeng (602–670).

4. Li Chunfeng's work on the *tianshu* (predestination) experts in history

In his preface to *Yisi Zhan* (*Yisi Astrology*), one of the two most celebrated astrological works in ancient China, Li Chunfeng (602–670) of the early Tang Dynasty classified the astrologists or astronomers in history into the following eleven categories:

(1) Founders of Chinese astrology or astronomy: Xuanyuan (alias Huang Di), Yao, Shun, Zhong, Li, Xi, and He, who initiated the study of astrology or astronomy, formed the group of folk astronomical customs, explored the principles in the books of Hetu and Luoshu, or were appointed as professional astrologists or astronomers.

(2) Professional astrologists or astronomers: Wuxian, Shi Shi, Gan Gong, Tang Mei, Zi Shen, and Bi Zao, who inherited astrological or astronomical studies from their founders or profoundly furthered their studies on the same.

---

[6] *Huainanzi Fanlunxun.*

(3) Expert astrologists or astronomers: Qi Zi and Zi Chan, who were well versed in the studies of astrology or astronomy in terms of its principle, literature, history, and system.

(4) Furthering astrologists or astronomers: Zhang Heng and Wang Xingyuan, who further explored the field, made astronomical apparatus and equipment, solved astronomical mysteries, or created their own theories.

(5) Scholarly astrologists or astronomers: Qiao Zhou, Guan Lu, Wu Fan, and Cui Hao, who did a thorough study of it in all aspects.

(6) Sacrificial astrologists or astronomers: Gu Yong, Liu Xiang, Jing Fang, and Lang Yizhi, who established a religion or worship of it, had it esteemed, sacrificed themselves for it, or made astrological suggestions to the royal court.

(7) Pseudo-astrologists or astronomers: Wang Shuo, Dangfang Shuo, Jiao Gong, Tang Du, Chen Zhuo, Liu Biao, and Xi Meng, who only half understood the concept, insisted on one theory or another, and left behind some short notes and lay sayings at legacies.

(8) Vernacular astrologists or astronomers: Han Yang and Qian Le, who sold their astrological or astronomical knowledge to the people on the street just to make a living.

(9) Comparing astrologists or astronomers: Cai Yong, Zu Geng, Sun Senghua, and Yu Jicai, who compared different theories, carried out special studies, made new discoveries, and filled in the missing blanks in the field.

(10) Evil astrologists or astronomers: Yuan Chong, who plagiarized others' studies, flattered the ancient experts, and persecuted benign astrologists.

(11) Persecuted astrologists or astronomers: Guo Pu, who was suspected and persecuted for his significantly great achievements and loyal suggestions in astrology.

It seems inappropriate to include Xuanyuan, Tang, and Yao in the first category, but this reflects the fact that kings were the leading shamans in the early Xia Dynasty. Witches and wizards, ranging from Wuxian to Bi Zao, also appeared in *Tianguan Shu of Shih Chi* (*Book of the Officials in Heaven, The Records of the Grand Historian*), but are not to be commented upon here. However, there is a conspicuous point in the aforementioned categories that will probably arrest the attention of readers: many people in the list are famous for other reasons than *tianxue*.

5. Important historical figures: Zi Chan, Ji Zi, Cai Yong, Liu Biao, and Zhuge Liang

In ancient times, people believed that astrology was an important subject with decisive influence on military or national decision-making. Despite banning common people's access to tianxue studies in various dynasties, high officials were allowed to master it as they had to share the weal and woe with the nation and take on heavy responsibilities.[7] Therefore, some historical figures, be it politicians or writers in modern people's eyes, were actually *tianxue* experts, and some even held specific positions like imperial astronomers. Let us look at some people on Li Chunfeng's list.

In the list provided by Li, a few people were not well known for *tianxue*, or did not hold any official posts in *tianxue*. Some of them are listed below.

Zi Chan. It is rather unconvincing for Zi Chan (?–552 B.C) to be on the list. As a great politician and diplomat in the Spring–Autumn Period, Zi Chan had worked as a minister for many years. Though it is not wrong to assume that he knew *tianxue* (to a certain extent), the historical affairs recorded in *Zuo Zhuan* (*The Commentary of Zuo*) showed his disbelief in subject of *tianxue*. In the 17th and the 18th year during the reign of Zhaogong (541 B.C.–510 B.C.), Bi Zao, the astrologist of Zheng State, foretold that a comet would bring fires to the states of Song, Wei, Chen, and Zheng. He suggested offering jade to the gods, but this suggestion was rejected by Zi Chan. The fire actually did target the four states in the end. With this prophecy fulfilled, Bi Zao foretold of another fire and offered the same advice. However, Zi Chan rejected it once more, regardless of the criticism he faced from his strongest supporter Zi Dashu. Zi Chan said:

The celestial laws are not reachable while the laws of the human world are around us. How can Bi Zao know the celestial laws? This is

---

[7] Jiang Xiaoyuan: *Tianxue Zhenyuan*, pp. 61–62.

because he makes bold predictions too often and it's nothing that he occasionally got his prophecies proved.

Though the ban was not lifted until the middle and late Ming Dynasty period, Emperor Renzong's attitude on this issue was almost the same as that of his predecessors.

Zi Chan directly pointed out Bi Zao's ignorance of celestial laws and said that his prophecies were merely coincidences. This was the severest way to attack an astrologist (Bi Zao might hold a life-long grudge against Zi Chan). Also, Zi Chan's famous remark about celestial laws was rarely seen in ancient China with the prevalent *tianxue*. Li Chunfeng thought highly of the both sides, however. In his view, Bi Zao "stuck to his opinion" and Zi Chan was commented as "knowing the cardinal principles" of celestial laws, despite his disbelief in it.

Ji Zi. Ji Zi once said to King Wu of Zhou (1087B.C.–1043 B.C.) that "Providence bestowed Yu nine natural laws for governance, which give the order to the conventional laws. ... There are five ages, namely, year, month, sun, star and calendar" (*Records of the Grand Historian: Hereditary House of Song Weizi*). Thus, Li Chunfeng (602–670) thought that he knew the cardinal principles.

Cai Yong. Cai Yong was a famous scholar in the Eastern Han Dynasty (25–220). In 178, the first year of the Guanghe Period, he and some senior officials including the imperial astronomer were called in by the Emperor. The Emperor asked him about calamities and ways to change and evade them. The story indicates that Cai was expert in astronomy and astrology. He authored the *Han Ji* (*The History of the Eastern Han*), which consisted of ten annals (Shi Yi). Among them, Annals 1 and 5 are about calendar and astronomy (*tianwen*).[8] *The History of the Sui Dynasty: Annals of Calendar*

---

[8]The following story suggests the meaning behind this. In the initial part of *Zhenze Changyu* by Wang Ao in the Ming Dynasty, he says:

> *One day, Emperor Renzong of Ming asked Yang Shiqi and other people: "Have you observed stars at night?" They replied that they knew nothing about them. The emperor said: "it's those who understand heaven, earth and people that are called real scholars, so why don't you know stars?" They replied: "the state decree stipulated that people aren't allowed to learn tianxue privately". The*

recorded that the court of the Eastern Han "ordered Liu Hong and Cai Yong to compile the calendar, with which Sima Biao wrote the sequel of the *History of the Eastern Han*". Later on, Cai was charged with a conviction and sent to garrison the border area in the north. However, he submitted a written statement to the court, recounting the past and recommending himself to the court, and this letter was recorded in *The History of the Later Han (Hou Han Shu): Annals of Calendar*. According to the *Older Book of the Tang Dynasty (Jiu Tang Shu): Annals of Calendar*, Liu Hong, Cai Yong, He Chengtian, Zu Chongzhi, and others "were all expert in arithmetic, but when they did calculations to compile a calendar, they were still whipped by far-fetched arguments". (see footnote 9) *The History of the Sui Dynasty: Astronomical Treatise* also stated that "Huan Tan, Zheng Xuan, Cai Yong and Lu Ji all give their own interpretations of *Zhou Bi Suan Jing*. But when checked against the actual astronomical phenomena, they were found to have committed fallacies". These historical records all revealed that Cai was an eminent astronomer in the late Eastern Han period.

Liu Biao. There is a book on astrology called *Jingzhou Zhan (Jingzhou Astrology)* authored by Liu Biao, an inconspicuous separatist in late Eastern Han, who resided in Jingzhou area. Hundreds of years later, during the Tang Dynasty, the book became an important reference for both Li Chunfeng's *Yisi Zhan (Yisi Astrology)* and Gautama Siddha's *Kai Yuan Zhan Jing (Astronomical Treatise of the Kaiyuan Era)*. In his book, Li Chunfeng listed 25 reference books on astrology authored by those that had known astrology by heart in their childhood, and Liu Biao's *Jingzhou Astrology* is listed in the 18th place. Liu's book was mostly produced by astrologists summoned by Liu Biao, who was the leader, just as many editors are doing today. Liu Biao was the link between the emperor and the subjects who may have been interested in astrology.

Another pertinent example can be cited here, which is quite amusing. *Tian Wen Zhi of Jin Shu (Astronomical Treatise, the Book of Jin)*,

---

emperor said: *"the decree is targeted among the people. You're the high-ranking officials of the state and your fates are closely linked to the fate of this state. How could you possibly not allowed to learn tianxue?"* Then the emperor granted them the right to learn tianxue.

which was also authored by Li Chunfeng, introduced the eight schools of astronomy in the chapter Astrological Locations of States and Prefectures. It made clear in the first place that the plan was unanimously adopted by the eight schools, which were represented by Chen Zhuo, Fan Li, Gui Guzi, Zhang Liang, Zhuge Liang, Qiao Zhou, Jing Fang, and Zhang Heng.

If we read modern works on astronomy, naturally we will be able to see that Chen Zhuo and Zhang Heng (78–139) both acted as the imperial astronomers in different dynasties. Chen Zhuo had even been the imperial astronomer during Eastern Wu, Western Jin, and Eastern Jin periods. He summed up the star Catalogues of three leading astrological schools. Zhang Heng authored the book *Ling Xian* (*Spiritual Constitution of the Universe*) and invented an apparatus that was capable of demonstrating the astronomical phenomena and forecasting earthquakes. Jing Fang was only mentioned in the footnotes. However, as for the remaining five people, there was no mention of them at all. But from Li Chunfeng's citations, we know that they authored astronomy books that had been passed on to later generations. Above all, Fan Li, Gui Guzi, and Zhang Liang could have written under other names.

6. Astrologists doomed by *tianxue*: Wu Fan, Geng Jicai, and Liu Ji

With a good mastery over celestial laws and provision of insightful counsel to the emperor, imperial astronomers used to enjoy high political status, which supposedly allowed these incumbents to demonstrate their desire for power and show off their talent. Nevertheless, joviality was less than a tag of being an imperial astronomer, and it could be called instead a far too perilous job, sometimes costing one's life! Let's first look at a few examples, the mechanism behind which will be better understood by retrospecting the previous sections.

Guo Pu. The name Guo Pu reminds scholars of *Ode to the Yangtze River* and *Poetry about Immortals*. He was thought of by Li Chunfeng as being the representative of *"mastered accomplishments court grudge and frank admonishment incurs death"*. As mentioned before, traditional Chinese *tianxue* was not aimed at exploration of the universe but was a means to

communicate with the heaven and earth — at its root, to provide governance counsel for the emperor to have a glimpse at the secret of future.

Therefore, an astronomer should not confine themselves to astronomy
and astrology, but should have also been an expert in many associated arts.
Guo Pu was just such a person. According to *Jin Shu (Book of Jin)*,
*"conversant at the five elements, astronomy and divination, Guo Pu knew
how to divert the misfortune, outwitting even Jing Fang and Guan Lu"*.
There are many wonderful tales about his capacity for foresight and his
clear knowledge of necromancy and astrology medicine scattered throughout his biography. For example, he foresaw what would happen long
before the chaos in Yongjia: *"He threw the book to the desk and exclaimed:
Alas! The people would be cornered by alien races and the farmland
would be trodden to wasteland"*. Then he planned to move to the southeast with his families and friends to evade the coming chaos.

Conversant at astronomy and taking part in court decision-making
were by all means honors, getting involved in treacherous politics was
quite another thing altogether. After moving to the south, Guo Pu was held
in high regard by Wang Dao, and won opportunities to show his future-
telling talents in front of the emperor, who was amazed by his talents.
However, wide acknowledgment for his talents led to him bringing disas-
ter to himself when Wang rose in revolt.

Before Qiao and Yu Liang suppressed Wang with armed force, they
came to Guo for his prediction and took home the answer of auspicious-
ness, which helped to further strengthen their confidence. However just
before his army set off, Wang Dun also came to Guo for divination. This
time, the answer "you would fail" and the admonishment for Wang to
return to Wuchang (not to revolt in Jiankang) for the sake of his own life
provoked Wang, who later beheaded Guo Pu. Hence, Li Chunfeng commented on this incident as *"frank admonishment incurring death"*.

Here, I thought I might tell the story of a Western astronomer who was
born more than 1,000 years later, as an amusing comparison with that of
Guo Pu. This story took place in 1610 as Germany was at civil war, when
Rudolph II (who later was reprimanded as insane) was then the emperor
of the Holy Roman Empire and Kepler, the astronomer and famous astrologer, was then the imperial mathematician at the court of Rudolph II. In
order to consolidate his ever-dwindling power, Rudolph II recruited a

mercenary army, and the opposition called in the King of Hungary, despite the fact that that he was in theory an official of the emperor. Both the parties wanted Kepler to practice divination for them. At this juncture, Kepler still stuck to his loyalty to the emperor despite of the predictable end of Rudolph II's dominance. On the one hand, he purposefully made an inauspicious prediction for the king, with an aim to impair his confidence; on the other hand, he admonished the adherents of the emperor that divination results should not be taken seriously when making great decisions, and said frankly that astrology *"should not only be expelled from the parliament, but from the heads of advisers for the emperor, and from the horizon of the emperor"*! However, his words were futile. The rebel army eventually stormed Prague and the emperor was made to abdicate.

In the above stories, both the astrologers had their own political stance and tried to influence the sequence of events. It might be safe to say that they adopted the same measures of action. But there comes a bigger problem: DO THEY TRULY BELIEVE THEIR PREDICTIONS? From the perspective of Kepler, he probably did not. As for Guo, he quite possibly shared the same mentality. That "saint uses spirit-like way to instruct" is the secret of ancient Chinese politics, which was well learned by politicians and sorcerers involved in politics, or they were well off the mainstream.

Wu Fan. Compared with Guo Pu, who ended up beheaded, Wu Fan's fate was a bit better. Wu Fan, styled Wen Ze, was born in Shangyu, Kuaiji (now Shaoxing, Zhejiang). He was known as an excellent master of fengshui and divination in Kuaiji. After Sun Quan rose to great power, Wu Fan came to serve in Sun's camp. Whenever there were good or bad omens, he could foresee them from divination most of the time. As a result, he became Sun's chief master of *tianxue*, responsible for astronomy and calendar matters. It is said that he predicted accurately about some major events during that period, including Liu Biao's death, the loss of Jingzhou, and Liu Bei's gain of "Yizhou" as well as Guan Yu's being defeated in Maicheng [*San Guo Zhi-Wu Shu* (*Book of Wu, Records of the Three Kingdoms*)].

However, Wu Fan and Sun Quan did not get along well with each other. Before Sun Quan became the king, Wu Fan once said to him, "there is *wang qi* in the south of the Yangtze River and a king is going to rise during 219–200 A.D.".. Such a statement was not only typical of masters of *tianxue* to inspire those politicians eyeing the throne and taking credit when those politicians succeeded, but a common method to mold public opinions and make propaganda for the political and military camp he served. Sun Quan then said to Wu Fan, "if your words turn out to be right, you will be ranked a noble". During 219–220 A.D., Sun Quan was made lord of Wu.[9] When he entertained his chancellors, Wu Fan reminded him of his promise, but he was only granted a ribbon. Furthermore, when Sun Quan awarded his chancellors according to their contributions, Wu Fan was specially excluded from the list of nobility he was placed originally in. Why did Sun Quan do this? Actually, Sun Quan himself was very interested in *tianxue* and wanted to learn from Wu Fan. However, Wu Fan wanted to keep his knowledge exclusive, so he did not tell Sun key points, which ignited Sun's resentment. However, though not conferred a rank of nobility, Wu Fan had a peaceful end. After he died from sickness, Sun missed him a lot and said, "If there were *tianxue* masters as great as Wu Fan and Zhao Da, I would like to make them Qianhuhou (a lord of 10,000 households). Unfortunately there is none".

Yu Jicai. Yu Jicai was a good example of one olden astronomer behaving properly and speculating well in times of political turmoil. He used to be the imperial astronomer of the Kingdom of Liang (502–557), but when the Kingdom of Zhou (557–581) seized Jiangling, he turned to serve for the latter. It was commonplace then and nobody would condemn him for lacking integrity or being a traitor as the Kingdom of Zhou was established by Xianbei, which was a foreign clan. According to *Sui Shu-Yishu Zhuan* (*Book of the Sui — Arts*), upon seeing Yu Jicai, Yuwen Tai treated him with great courtesy and made him an imperial astronomer. Whenever there was a war, Yuwen Tai would take him by his side. In the second year of the reign of Emperor Wucheng, Yu Jicai, together with Yu Xin and Wang Bao,

---

[9] The year 219 A.D. is the 24th year of the Jian'an Period (*196–219*) and the following year the first year of the Yankang Period of the Eastern Han Dynasty and also the year Cao Pi took the throne of the Wei Dynasty. But San Guo Zhi (*The History of the Three Kingdoms*) *still took it as the 25th year of the Jian'an Period (196–219). It was the Genzi Year. So what Wu Fan referred to is these two years.*

both of whom were chancellors of the Kingdom of Liang, became the Linzhi Scholars of the Kingdom of Zhou.

During the 10-plus years of the reign of Emperor Yuwen Yong, the state power fell in the hands of Yuwen Hu, who was regarded as the actual ruler of the Kingdom of Zhou. When Yuwen Yong killed Yuwen Hu, he, without doubt, killed all those chancellors who in the past had curried favor with Yuwen Hu. Yuwen Yong checked all the letters and documents confiscated from Yuwen Hu and all those who had incited Yuwen Hu to rise in rebellion were killed using the excuse of heaven's will. Only Yu Jicai, in his two letters, tried to persuade Yuwen Hu to give the power back to Yuwen Yong by telling him about celestial phenomena and signs of calamity. Consequently, Yuwen Yong thought quite highly of Yu Jicai, regarded him as an honest and dutiful chancellor, and gave him a promotion.

Though Yu Jicai won the trust of Yuwen Yong for his ingenious handling of this matter, being in the king's company is tantamount to living with a tiger. Since the start of the Sui Dynasty (581–618), the strife among masters of *tianxue* had become fierce. Yu Jicai was finally deposed for speaking frankly.

This was a time when emperor Wen of the Sui Dynasty deposed Prince Yang Yong to the rank of a common person and appointed Yang Guang (the latter Emperor Yang of the Sui Dynasty who later subjugated the state) as his heir. The then imperial astronomer Yuan Chong always flattered the emperor by saying auspicious words, and this time he said, *"Since the start of the Sui Dynasty, the day is getting longer. Such phenomenon is rare since the ancient times. The reason must be that the heaven has been moved by the good reign initiated by Sui"*. Though in ancient China, a great many people believed in the interaction between heaven and mankind, it was the height of ridiculousness to say that the day would get longer for the good reign of the Sui. Yuan Chong's fallacy was sneered at by the fellow imperial tianxue officials and later generations, but it fitted in exactly with the wishes of Emperor Wen, who later went to ask for Yu Jicai's opinion. Yu, however, criticized Yuan for his ridiculous remarks, which angered the emperor, who finally removed him from office and sent him back to his home place with his emolument cut by half. However, after that, the emperor often sent people to his home whenever there was any positive or negative omen. In this way, Yu brought his afterheat into play.

Yu died peacefully at the age of 88, which was not a bad end at all.

Liu Ji. The last example is Liu Ji, from which we can see from Liu that being the emperor's advisor is like wading in deep water and treading on thin ice. Liu Ji used to be a *jinshi* (a successful candidate in the highest imperial examinations) in the Yuan Dynasty (1271–1368) and had held office several times. But he abandoned his official career to live in seclusion because of the corrupt court. Later, he was recruited by Zhu Yuanzhang and became one of Zhu's top advisors. He helped Zhu ascend to the throne. As Zhu's top *tianxue* master, he was made imperial astronomer after Zhu came to the throne. He introduced *Wu Shen Da Tong Li* (a kind of astronomical almanac), the adaptation of *Shou Shi Li (Shoushi Calendar)* by Guo Shoujing in the Yuan Dynasty. The almanac had been used till the end of the Ming Dynasty.

When performing the duties of an imperial *tianxue* master, Liu Ji was very cautious. According to *History of the Ming Dynasty: Biography of Liu Ji*, the emperor once wrote a confidential letter to Liu Ji, asking for explanations of astronomical phenomena. Liu Ji elaborated them one by one and then burned his manuscript (to keep it confidential). After Zhu Yuanzhang came to the throne, Liu should have been rewarded with a high position and great wealth as he was one of the founding fathers of the dynasty. Knowing that "after the cunning hare is killed, the hound will be boiled; when the birds are gone, the bows are shelved", he wanted to escape from being killed by Zhu Yuanzhang, who would definitely execute those who had helped him. As a result, he secluded himself in mountains. He would drink and play Weiqi, refraining from bragging about his feats. Nevertheless, he did not succeed in evading his fatal destiny.

Hu Weiyong, Liu's political opponent, reported to the throne that Liu attempted to keep the land with "*wang qi*" (bearing, air, or manner typical of kings and emperors) and use it as his graveyard. Such an accusation was no trivial matter.

According to relevant theories in ancient Chinese *tianxue*, where there is *wang qi* (bearing, air, or manner typical of kings and emperors), there will rise an emperor. The rise of great powers like the Han Dynasty and the Tang Dynasty certainly had the *wang qi* as omen. Small dynasties like

Chen in the east of the Yangtze River also had *wang qi*. After it was destroyed by Emperor Wen, its *wang qi* dissipated. Even the puppet regimes of Chu and Qi, which existed only for a short time, also had the *wang qi*. Let us take Volume 8 of Ting Shi (*The History of Ting*) — *Wang Qi in Fucheng* as an example.

During the reign of Emperor Huizong of the Song Dynasty, it was reported to the throne that there was obvious wang qi in Fucheng but Emperor Huizong did not believe that. Soon many alchemists reported the same thing and the emperor issued an imperial edict to destroy mountains around Zhilong to clear up the wang qi. But after one year, it was still there, a bit gloomy though. As a result, the wang qi there was regarded as sign of an unorthodoxy regime. Till the reign of Emperor Qingzong, the Chu puppet regime was established but collapsed within one month. After Liu Yu rebelled and established the puppet regime Song, he made "Fuchang" the regime title and asked the Jin for labourers to restore those places. The workers toiled all year round and could not stand it, which naturally led to the overthrow of the puppet regime. The two puppet emperors Zhang Bangchang and Liu Yu all came from Fucheng, which was consistent with the divination.

Such being the case, if Liu Ji had attempted to have a finger in the pie that was *wang qi*, he must have been eyeing the throne. When Liu Ji helped Zhu Yuanzhang to the throne, he had probably heard of such kind of sayings like "*Wang Qi* in Fengyang (hometown of Zhu Yuanzhang)" or "*Wang Qi* in Jinling (the present Nanking)". Many founding emperors would worry that since imperial masters of *tianxue* could help them to the throne, they could also take the throne from their heirs. Hu Weiyong took advantage of this thought group to beat and insult Liu Ji. Though Zhu Yuanzhang did not believe the whole story, he still deprived Liu Ji of his official salary. Terribly scared, Liu Ji came to the capital to apologize and dared not go back as those were times when if the emperor said you were guilty, you must be guilty (whether you were actually guilty or not).

Worried and indignant, Liu Ji soon got sick. After having the medicine prescribed by the doctor sent by Hu Weiyong, he was in a worse condition. It was said that Liu Ji was poisoned to death. On his deathbed, he made another effort to keep his descendants from calamity, as is described below.

When dying, Liu Ji gave the Tian Wen Shu (Book on Astronomy) to his son Lian and said, "Hand in the book to the emperor as soon as possible and my descendents should never learn it!"[10]

These words became the last words of a once well-known great master of *tianxue*.

---

[10]We will mention in the next chapter that imperial masters of *tianxue* are usually hereditary, especially in the Ming Dynasty. As Liu Ji was the former imperial astronomer, his son was rightful to learn astronomy.

Chapter 4

Government-run *Tianxue*: Tradition and Exception

1. *The Rites of Zhou* and offices for *tianxue* after the Zhou Dynasty

For thousands of years, the imperial organization for *tianxue* had been a major department of the "central government" in ancient Chinese society. This tradition can be dated back to the Zhou Dynasty or even earlier. According to *Canon of Yao, Shangshu*, Emperor Yao sent *tianxue* officials to different regions of the country to observe celestial phenomena, from which we can infer that there was perhaps an office for *tianxue* even before the Zhou Dynasty. Considering the fact that the scholars of this office were also important officials, we can say that such an assumption was not far-fetched. In the Zhou Dynasty, the imperial organization for *tianxue* was much larger, staffed with more officials in this regard, and exerted a profound and far-reaching impact. *Zhou Li* (*The Rites of Zhou*) provides some insight to the imperial organization for *tianxue* at that time.

There has been controversy with regard to when the book *The Rites of Zhou* was produced. It was even denounced as a pseudograph when skepticism of the ancient period prevailed. However, according to recent research, the official system of the Western Zhou Dynasty, as revealed by bronze inscriptions, stated something similar to what was said in the *The Rites of Zhou*. So it can be concluded that the official system recorded in The Rites of Zhou is mostly the record of the actual official system of the Western Zhou,[1] though there could be some elements visualized by later generations.

In *Offices of Spring* (*Chunguan Zongbo*), *The Rites of Zhou* enumerated at least six offices whose duties were related to *tianxue*.

Da Zong Bo was an office in charge of sacrificial rites for gods in the heaven and on the earth as well as for ghosts to assist the king in stabilizing the country. It was responsible for burning sacrificial matter on wood to worship the sun, the moon, and stars.

Zhan Meng was responsible for the studying of seasons, observing the sky and earth, discerning *yin* and *yang*, and reading the six dreams according to the sun, the moon, and stars.

Shi Jin was responsible for judging fortunate and unfortunate omens by observing the ten colors of the sunshine, so as to tell if the future is fortunate.

Da Shi (Tai Shi) made adjustments to the calendar so as to keep the calendar year synchronized with the astronomical year, planned events for the people of the country for each season, and issued the plan to the offices and benefices. It also issued the calendar policy for the next year to all principalities. During leap months, the office asked the king to stay in the palace for the entirety of the month. During large sacrificial ceremonies, Da Shi practiced divination and worshipped the sun with other officials.

Feng Xiang Shi was in charge of observing the *Taisui* (ancient divine name of Saturn), which spent twelve years moving round the heaven once, the moon that waxed and waned 12 times every year, the twelve stars that the handle of the Dipper pointed to, the ten days of one-third of every month, and the twenty-eight lunar mansions so as to make calendars. It measured the length of the shadow cast by the sun in winter and summer and measured the shadow cast by the moon in spring and autumn with a gnomon, so as to determine the order of the seasons.

Bao Zhang Shi studied the movements of the sun, the moon, and stars to learn the movements in the world and to foretell if these movements are fortunate. It decided on the territory of the country according to the demarcation of the mansions, as every benefice is governed by a star which shows good or bad omens. It distinguished between fortunate and unfortunate omens by observing the twelve zodiacs. It foretold whether

---

[1] Chen Hanping, *Study on the Office System of the Western Zhou Dynasty*; Xuelin Press, (1986), p. 214.

the future was fortunate or unfortunate and also forecasted flood, drought, harvest, and famine according to the five colors of the cloud. The office also learnt whether the relation between the heaven and the earth was harmonious and discerned the related omens by observing the wind of the twelve months. Through these five ways of divination, Bao Zhang Shi would then advise the King to make remedies to his governing strategy and make arrangements for things that should be done.

The title and the number of the staff of these offices were also clearly defined.

Da Zong Bo had only a Qin.

Zhan Meng had eight subordinate officials, including two Zhong Shis, two Shis, and four Tus.

Shi Jin had two Zhong Shis, two Shis, and four Tus under its leadership.

Da Shi had six subordinate officials, including two Xia Dafus and four Shang Shis.

Feng Xiang Shi had two Zhong Shis, four Xia Shis, two Fus, four Shis, and eight Tus under it.

Da Zong Bo was in charge of a number of affairs, and *tianxue* was just one of them. The grade of Da Shi (Tai Shi) was also very high, next only to Da Zong Bo, and Da Shi also had many other duties besides *tianxue*. The other four offices were of lower grades and responsible for more trivial affairs.

There is no doubt that offices recorded in *The Rites of Zhou* had a profound influence on the administrative structure of ensuing Chinese governments. The six first-class offices evolved into the six ministries of the central government, with the offices of spring becoming the Ministry of Rites to which the imperial organizations for *tianxue* had been a subordinate for two thousand years.

The name and title of the head official of the imperial organization for *tianxue* changed many times, as the table below shows.[2]

---

[2] Referred to *"Tong Dian"* (Vol. 26), *"Xu Tong Zhi"* (Vol. 131) and *"Xu Wen Xian Tong Kao"* (Vol. 56). Also referred to Wang Baojuan: "Astronomical Institutions in the Tang Dynasty", in A Collection of Chinese Astronomical History, Science Press, (Issue 5) (1989); Wang Baojuan: "Astronomical Institutions in the Song Dynasty", in *"Chinese Astronomical History"*, Science Press, (Issue 6) (1994); Wang Baojuan: "Astronomical Institutions in the Liao, Jin and Yuan Dynasties", "A Collection of Chinese Astronomical History", Science Press, (Issue 6) (1994).

Office	Head of the Office	Era
	Taishi Ling (Court Astronomer)	Qin Dynasty (221–207 B.C.)
	Taishi Gong (Grand Historian)	the reign of Emperor Wu of Han (141–87 B.C.)
	Taishi Ling (Court Astronomer)	from the reign of the Emperor Xuan of Han (74–49 B.C.) to the Northern and Southern Dynasties
Taishi Cao (Bureau of Astronomy)	Taishi Ling (Court Astronomer)	the reign of the Emperor Wen of Sui (581–604)
Taishi Jian	Taishi Ling	the reign of the Emperor Yang of Sui (604–618)
Taishi Ju	Taishi Ling	early Tang Dynasty
Mishu Ge	Langzhong	the second year of Longshuo Era of the reign of the Emperor Gaozong of the Tang Dynasty (662)
Taishi Ju	Taishi Ling	the first year of Xianheng Era of the reign of the Emperor Gaozong (670 A.D.)
Huntian Jian	Huntian Jian	the first year of Jiushi Era of the reign of the Empress Wu Zetian (700 A.D.)
Hunyi Jian	Hunyi Jian (700AD)	the first year of Jiushi Era of the reign of the Empress Wu Zetian
Taishi Ju	Taishiju Ling	the second year of Chang'an Era of the reign of the Empress Wu Zetian (702 A.D.)

Taishi Jian	Taishiju Ling	the second year of Jinglong Era of the reign of the Emperor Zhongzong of Tang (708 A.D.)
Taishi Jian	Taishi Jian	the second year of Kaiyuan Era of the reign of the Emperor Xuanzong of Tang (714 A.D.)
Taishi Ju	Taishi Ling	the 14th year of Kaiyuan Era of the reign of the Emperor Xuanzong of Tang (726 A.D.)
Taishi Jian	Da Jian	the first year of Tianbao Era of the reign of the Emperor Xuanzong of Tang (742 A.D.)
Sitian Tai	Da Jian	the first year of Qianyuan Era of the reign of the Emperor Suzong of Tang (758 A.D.)
Sitian Jian	Sitian Jian	from the early Northern Song Dynasty to the reign of Emperor Shen of Song (1067–1085)
Taishi Ju	Taishi Ling	from the reign of Emperor Shen of Song (1067–1085) to the late Southern Song (1127–1279)
Sitian Jian	Taishi Ling	Liao Dynasty (907–1125)
Sitian Jian	Ti Dian	Jin Dynasty (1115–1234)
Taishi Yuan	Yuan Shi	Yuan Dynasty (1271–1368)
Sitian Tai	Sitian Jian[3]	Yuan Dynasty (1271–1368)
Qin Tian Jian	Jian Zheng	Ming Dynasty (1368–1644)
Qin Tian Jian	Jian Zheng	Qing Dynasty (1636–1912)

---

[3] The imperial organization for *tianxue* in the Yuan Dynasty was different from those of other dynasties. For instance, Taishi Yuan was staffed with five Yuan Shis; Sitian Tai had three Sitian Jians and these three Sitian Jians were subordinate to Ti Dian (this may be a nominal leader).

Despite the different names, these imperial organizations for *tianxue* had the same status and duties as the offices related to *tianxue* recorded in *The Rites of Zhou*, and such a status and duties stayed almost the same for thousands of years.

2. Grades of the officials who worked in *tianxue*

In *Mathematics and the Sciences of the Heavens and Earth*, in the 3rd volume of *Science and Civilization in China*, Joseph Needham quoted the following exaggerated remarks on ancient Chinese calendar policy said by Richard F. Kuhnert in the 19th century:

Probably another reason why many Europeans consider the Chinese such barbarians is on account of the support they give to their Astronomers-people regarded by our cultivated Western mortals as completely useless. Yet there they rank with Heads of Departments and Secretaries of State. What frightful barbarism![4]

Kuhnert emphasized the high status of the ancient Chinese astronomers. But his understanding was not very accurate. The imperial organizations for *tianxue* were always under the charge of the Ministry of Rites, whose leader and leaders of the other five ministries were "Heads of Departments of the State". If an astronomer took higher ranking, that was not because his role was that of an astronomer, and the position of Taishi Ling (the Court Astronomer) was held by someone else and not this high-position astronomer.

The rank of the head of the imperial organizations for *tianxue* varied with the dynasty, which can be seen from the table below (whose data has not been completed yet).

If we are to find a corresponding official for Jian Zheng of Qintianjian of the Ming and Qing Dynasties in modern office rank system, it may be an official lower than vice-minister and higher than director-general — although Qin Tian Jian was subordinate to the Ministry of Rites, it was like an "institution directly under the State Council". Today, China's five Astronomical Observatories are all administrated by Chinese Academy of Sciences. The head of the Observatory is of director-general level while

---

[4] F. Kuhnert: Das Kalenderwesen bei d. Chinesen, Osterreichische Monatschrift f. d. Orient, Vol. 14 (1888), citied from Joseph Needham. *Science and Civilization in China* (*Vol. 4*). Translated version, Science Press, (1975), p. 2.

the head of the Academy of Sciences is of the ministerial level. However, there was only one Qin Tian Jian in the government and what this office was engaged in was exclusively griped by the imperial court, it is reasonable that we say the grade of Jian Zheng should be slightly higher than the head of the Astronomical Observatory today.

Head	Official Grade	Era
Taishiju Ling	Lower Deputy Fifth Grade	the fourth year of Wude Era of the reign of Emperor Gaozu of Tang (621)
Da Jian	Deputy Third Grade	the first year of Tianbao Era of the reign of Emperor Xuanzong of Tang (742)
Da Jian	Deputy Third Grade	the first year of Qianyuan Era of the reign of Emperor Suzong of Tang (758)
Sitian Jian	Regular Third Grade	early Northern Song Dynasty
Ti Dian of Sitian Tai	Regular Fifth Grade	Jin Dynasty
Yuan Shi of Taishi Yuan	Regular Second Grade	Yuan Dynasty
Sitian Tai	Deputy Fifth Grade	Yuan Dynasty
Jian Zheng of Qin Tian Jian	Regular Second Grade	Ming Dynasty
Jian Zheng of Qin Tian Jian	Regular Second Grade[5]	Qing Dynasty

Now, let us review what Kuhnert said, and thus we can find that he overestimated the level of the head of imperial organizations for *tianxue*.

---

[5] According to Volume 35 of *Li Dai Zhi Guan Biao* (*Table of Offices of Previous Dynasties*) compiled by Ji Yun, Jian Zheng of Qin Tian Jian ranked Fourth Grade in early Qing Dynasty, then started ranking Third Grade in the sixth year of Emperor Kangxi (1667), and the Western and Manchurian Jian Zhengs had ranked Regular Fifth Grade since the ninth year of Emperor Kangxi (1670).

Sometimes, the ancient officials for *tianxue* were in the favor of the emperor and enjoyed status as the emperor's teacher, which was a level much higher than "Secretaries of State". This is beyond Kuhnert's comprehension.

3. The evolution of the imperial organization for *tianxue*

Imperial *tianxue* organizations in different dynasties varied in size.

The number of the staff in each imperial organization for *tianxue* in the Zhou Dynasty was about forty, according to *the Rites of Zhou*. But this figure may not be true.

According to *Xi Han Huiyao* (*References from Western-Han-era Histories*) compiled by Xu Tianlin, a scholar in the Song Dynasty, some information from *Treatise on Rhythm and the Calendar, Treatise on Sacrifices,* and *Li Guang* of the *Han Shu* (*Book of Han*) suggested that in the Western Han Dynasty there had been Da Dianxing, Zhi Li, Wang Qi, Wangqi Zuo, and some other offices for *tianxue*.[6] This References book included more information on the imperial organization for *tianxue* in the Eastern Han Dynasty, but the office we know existed for sure is the Taishi Ling. Taishi Ling received a salary of 600 *dan* of grain (this level of salary was similar to that of the Taishiju Ling in the Tang Dynasty, who ranked Lower Deputy Fifth Grade). Taishi Ling had three subordinates, one Cheng, one Mingtai Cheng, and one Lingtai Cheng, each of whom only received a salary of 200 *dan* of grain.[7] However, Liu Zhao's *Hou Han Shu Zhi Zhu* (*Annotations to the Book of Later Han*) quoted the information of Taishi Ling from *Han Guan Yi* (*Official Rites of the Han Dynasty*) and stated that Taishi Ling had thirty-seven officials under its direct leadership, and Lingtai Cheng, a subordinate office of Taishi Ling, had forty-two staff besides the head.[8]

In the following, we list the division of duties of the thirty-seven staff members directly under Taishi Ling:

Zhi Li, six
Gui Bu, three

---

[6] Xu Tianlin, Volume 31 of *Xi Han Hui Yao* (*References from Western-Han-era Histories*).
[7] Xu Tianlin, Vol. 19 of *Dong Han Hui Yao* (*References from Western-Han-era Histories*).
[8] Quoted from Vol. 35 of *Li Dai Zhi Guan Biao* (*Table of Offices of Previous Dynasties*).

Lu Zhai, three
Ri Shi, four
Yi Shi, three
Dian Rang, two
Ji Shi, three
Xu Shi, three
Dian Chang Shi, three
Jia Fa, two
Qing Yu, two
Jie Shi, two
Yi, one

The following is the division of duties of the forty-two staff of Lingtai Cheng:

Hou Xing, fourteen
Hou Ri, two
Hou Feng, three
Hou Qi, twelve
Hou Guijing, three
Hou Zhonglü, seven
She Ren, one

We can infer the size of the imperial organization for *tianxue* in the Han Dynasty from this information.

There was little information on the imperial organizations for *tianxue* in the Three Kingdoms, Western and Eastern Jin Dynasties, and the Northern and Southern Dynasties. Ji Yun, a scholar in the Qing Dynasty, compiled *Li Dai Guan Zhi Biao* (*Table of Offices of Previous Dynasties*) and quoted the following sentence from *Tang Liu Dian* (*Six Codes of Tang Dynasty*), "The Taishi Ling of the Kingdom of Wei was staffed with twenty Wanghou Langs and fifteen Houbu Langs", "Tai Shi (of the Jin Dynasty) had four Dian Lis, twenty Wanghou Langs and fifteen Houbu Langs under its leadership". As *Baiguan Zhi* (*Government Offices*) of *Sui Shu* (*Book of Sui*) stated, "Taishi Ling during the reign of the Emperor Wen of Sui had four Sichen Shis and one hundred and ten Louke Shengs under it".

The historical records of the Tang Dynasty and ensuing dynasties provide an abundance of information in this regard. Some scholars have carried out studies on the imperial organizations for *tianxue* in the Tang, Song, Liao, Jin, and Yuan dynasties. Details are given in the following.

The imperial organization for *tianxue* in the Tang Dynasty was large. The size of the organizations for *tianxue* reached the peak during the first year of the Qianyuan Era during the reign of the Emperor Suzong of Tang (758 A.D.) when Taishi Jian changed its name into Sitian Tai, and another bureau for *tianxue* was established with new personnel. Details are shown below, which provides insight to the organization for *tianxue* in the Tang Dynasty.

Da Jian, one
Shao Jian, two
Wuguan Baozhang Zheng, five
Cheng, three
Zhu Bu, three
Ding E Zhi, five
Wuguan Lingtai Lang, five
Wuguan Sili, five
Wuguan Jianhou, five
Wuguan Qiehu Zheng, five
Wuguan Sichen, fifteen
Wuguan Lisheng, fifteen
Wuguan Kaishushou, five
Ling Shi, five
Louke Boshi, twenty
Dianzhong, Diangu, three hundred and fifty
Tianwen Guansheng, ninety
Tianwen Sheng, fifty
Li Sheng, fifty-five
Lou Sheng, forty
Shi Pin, ten

The total number of staff under the office was 694.[9] Few organizations could compete with such a big organization for *tianxue* in terms of size in

---

[9]The transmission and influence of Western astronomy in the Ming and Qing *dynasties* in China will be discussed in details in later chapters.

the whole world. The office was later downsized, but the number of staff still was at 671. But Dian Zhong and Dian Gu may be guards of honor.

The imperial organizations for *tianxue* in the Liao, Jin, and Yuan dynasties were of smaller sizes, with around one hundred subordinate offices.[10]

The Yuan Dynasty inherited the organizations and officials for *tianxue* of the Song and Jin *dynasties*. Besides, a Huihui Sitian Tai (for Hui nationality) was established in Shangdu (southeast of today's Duolun County of Inner Mongolia), another Sitian Tai was set up in Yuandadu (currently Beijing) as the office of Taishi Yuan, and a Guanxing Tai (star observatory) was also set up in Yangcheng (currently Dengfeng, He'nan Province). Large-scale celestial measurements were also conducted in the whole nation. The organizations for *tianxue* in the Yuan Dynasty were large and complicated. Yuan Shi, the head of the Taishi Yuan, the imperial organization for *tianxue* in early Yuan Dynasty, ranked in Regular Second Grade, Sitian Tai and Huihui Sitian Jian were similar to Taishi Yuan (their heads were of lower level than that of the Taishi Yuan). In the tenth year of the Yuan Dynasty (1279), a Sitian Tai was established in Yuandadu with Tuisuan Ju, Ceyan Ju, and Louke Ju under it. These three Jus (bureaus) were staffed with 70 people. In the first year of Huangqing Era (1312), Sitian Jian had 120 subordinate offices.[11]

4. The imperial organizations for *tianxue* in Ming and Qing dynasties

In the Ming and Qing dynasties, the centralized and autocratic monarchical system was highly developed. The sovereign power relied less on imperial organizations for *tianxue,* but such organizations still maintained their high status. The organization for *tianxue* in the Ming and Qing

[10] Here, the "Han Soldier/Army" and the "Han People" are very different: the "Han Army" consist of soldiers of the Eight Banners. The Eight Banners of the Qing Dynasty consists of the Manchu Eight Banners, the Mongolian Eight Banners and the Han Army Eight Banners. The "Han Army" joined the conquerors after being conquered; the "Han Army" are true conquered people.

[11] See Volume 35 of *Lidai Guanzhi Biao* (*Table of Offices of Previous Dynasties*) and the third essay of the collection of Wang Baojuan.

dynasties was named Qin Tian Jian, which was used for over half a century and became a popular word in ancient Chinese dramas and novels.

According to *Zhiguan Zhi* (*Government Offices*), *Ming Shi* (*History of Ming*), Qin Tian Jian of the Ming Dynasty was of relatively small size with the following staff:

Jian Zheng, one
Jian Fu, two
Zhu Bu of Zhubu Ting, one
Chun Guan Zheng, Xia Guan Zheng, Zhong Guan Zheng, Qiu Guan Zheng, Dong Guan Zheng, one each
Wuguan Lingtai Lang, eight
Wuguan Baozhangzheng, two
Wuguan Qiehuzheng, two
Wuguan Jianhou, three
Wuguan Sili, two
Wuguan Sichen, eight
Louke Boshi, six

The total number of staff was 40, and the number was decreased to 22 later. Jian Zheng was the head of the organization, ranking Regular Fifth Grade, and Wuguan Sichen, and Louke Boshi were Deputy Ninth Grade. Qin Tian Jian was divided into four departments.

Tianwen Ke (Department of Celestial Phenomena), which was responsible for observing and recording celestial phenomena.

Louke Ke (Department of Hourglass), which was responsible for synchronizing time.

Li Ke (Department of Calendar), which was responsible for making *Da Tong Li* (*Datong Calendar*).

Huihui Ke (Department of Hui Nationality), which evolved from the Huihui Sitianjian of the Yuan Dynasty and early Ming Dynasty and was engaged in Islamist *tianxue* and provided supplement and references for traditional Chinese *tianxue* with Islamist methods for studying *tianxue*.

Compared with previous dynasties, Qin Tian Jian of the Ming Dynasty was really small. This was associated with the fact that *tianxue* was less important and became an accessory for sovereign power and also because the restriction on ordinary people studying *tianxue* was gradually relaxed.

In the late Ming Dynasty, however, another two organizations for *tianxue* were established. *Da Tong Li* was used for a long time, and the drift of this calendar became more obvious. At that time, many Jesuits came to China and presented advanced Western astronomical approaches. As a result, there was an ardent call for revision of the calendar inside and outside the court. But Qin Tian Jian cleaved to old approaches and strongly detested such appeal. In the second year of the Chongzhen Period, Li Ju (Bureau of Calendar) was set up with Xu Guangqi as its head to compile *Chongzhen Lishu* (*Chongzhen Almanac*). As Li Ju focused its work on introducing Western astronomy, it was also called "Xi Ju" (Western Bureau). Wei Wenkui, a man without any title, was opposed to the introduction of Western astronomy. The group of people led by Wei, with the support of some high-ranking officials, was called "Dong Ju" (Oriental Bureau). *Li Zhi* (*Calendar*) of *Ming Shi* (*History of Ming*) stated that (see footnote 9):

At that time, four organizations were engaged in calendar making — Datong, Huihui, Xi Ju introducing Western approaches and Dong Ju led by Wenkui. Their theories were different and no agreement was reached between them.

The argument between these "official" organizations for *tianxue* was unprecedentedly heated. In the Yuan Dynasty, Huihui Sitian Tai and Sitian Tai of Han competed with each other, and complemented each other, but they were not opposing each other. Dong Ju and Xi Ju were both temporary organizations that gradually disappeared as the Ming Dynasty was overthrown.

Qin Tian Jian (The Imperial Board of Astronomy) in the Qing Dynasty was different from that of the Ming Dynasty in two aspects. First, Emperor Shunzhi appointed Johann Adam Schall von Bell, a Western missionary, as the director of Qintian Jian. Thus, the tradition of appointing Western Jesuits to be directors of Qintian Jian took shape, and this tradition lasted for the next two hundred years. Second, as Manchu was an "intruder" of China, the government was very sensitive to the issue of nationality, and the administrative structure consisted of two groups, a group of Hans and a group of Manchurians. Therefore, the size of Qintian Jian in the Qing Dynasty was bigger than that of the Ming Dynasty.

Qintian Jian in the Qing Dynasty consisted of four departments, including Shixian Ke (Department of Calendar), Tianwen Ke (Department of Celestial Phenomena), Louke Ke (Department of Hourglass), and Zhubu Ting (Department of Files). *Li Dai Guan Zhi Biao* (*Table of Offices of Previous Dynasties*) stated that:

Qintianjian (The Imperial Board of Astronomy)

Jian Zheng, a Manchurian, and a Westerner

Jian Fu, a Manchurian, and a Han

Zuo Jian Fu, You Jian Fu, and a Westerner

Zongli Jianwuwang Minister, one (specially set up in the 15th year of Emperor Qianlong, and the number of personnel in this office was not fixed)

Shixian Ke:

Wuguan Zheng, two Manchurians and two Mongolians

Chun Guan Zheng, Xia Guan Zheng, Zhong Guan Zheng, Qiu Guan Zheng, Dong Guan Zheng, and a Han

Qiuguan Zheng, a Han soldier (see footnote 10)

Wuguan Sishu, a Han

Bo Shi, a Manchurian, two Han soldiers, two Mongolians, and sixteen Hans

Tianwen Ke:

Wuguan Lingtai Lang, two Manchurians, a Mongolian, a Han soldier, and four Hans

Wuguan Jianhou, a Han

Bo Shi, three Manchurians, and a Han

Louke Ke:

Wuguan Qiehu Zheng, a Manchurian, a Mongolian, and two Hans

Wuguan Sichen, a Han soldier

Boshi, six Hans

Zhubu Ting:

Zhubu, a Manchurian and a Han

Assistants:

Shifeng Tianwen Sheng, sixteen Manchurians, sixteen Mongolians, eight Han soldiers, and twenty-four Hans

Shiliang Tianwen Sheng, sixteen Hans

Shiliang Yinyang Sheng, ten Hans

Bitie Shi, eleven Manchurians, four Mongolians, and two Han soldiers

The total number of staff reached 196.

5. *Yin* and *Yang* system in different places

The capital-based Imperial *Tianxue* Institute had been the one doing *Tianxue* studies until the arrival of Yuan and Ming Dynasties when a pro-*Tianxue* institute was set up, what came to be known as the Yin and Yang System, the two opposing elements in nature.[12]

As was proved in my work *Tianxue Zhenyuan (The Truth of the Sciences of the Heaven)*, *Tianxue* is what underpins Yin and Yang divination in ancient China.[13] As such, it makes perfect sense that the *Tianxue* Institute saw Yin and Yang divination as what must be learned and used.[14] That said, backed by a profound foundation, the subject could not be easily manipulated by the royalty. In order to master and control it, the royal institute set up the Yin and Yang system.

The Yin and Yang system was established in 1291. According to *Yuanshi Xuanjuzhi Volume One (Volume One of Electoral Record of History of the Yuan Dynasty)*:

Kublai Kahn, Emperor Shizu of the Yuan Dynasty started to build a strong pool of Yin and Yang experts in June, 1291. The emperor took in those who excel in the subject from Fuli, north and west of Taihang Mountains, and Jiangnan, south of Yangtze River Delta. He gave those punters professional titles so that they could teach the science. Anyone with the talent could register themselves to provincial governments and later went to the capital for an examination. The truly talented would be appointed by Sitiantai, a government body. Soon after 1314 when Yanyou, emperor Renzong of Yuan Dynasty, reigned, the emperor gave them the same privileges as Confucian physicians, who were highly regarded then.

---

[12] See Shen Jiandong: "A Preliminary Study on the Yin and Yang System of Yuan and Ming Dynasties", Mainland Magazine (Taiwan, China), Vol. 79, No. 6, (1989). This section is mainly referred to the above results of Shen.

[13] See Jiang Xiaoyuan: *Tianxue Zhenyuan (The Truth of the Sciences of the Heaven)*, (2007) pp. 46–55.

[14] This point shall be discussed further in following sections.

Yin and Yang experts were appointed as teachers and governors at Lu, Fu and Zhou levels (administrative regions below the provincial level). They were governed by Taishi, a higher rank.

It can be seen from the record that non-governmental Yin and Yang gurus were placed under the jurisdiction of the government, and they were likely to be elected as back-up for the Imperial *Tianxue* Institute. The Ming Dynasty saw a more full-fledged Yin and Yang system with the establishment of Yin and Yang Zhengshu at the Fu level, Dianshu at the Zhou level, and Xunshu at the county level. These newly set up official ranks were so insignificant that Zhenshu was the ninth rank in the government without salary while Dianshu and Xunshu were completely off rank. Local officials dedicated to Yin and Yang were selected by the Imperial Board of Astronomy. According to Volume Two, Three, and Four of *Daming Huidian (Code of Great Ming Dynasty)*, those who were recommended to the position must sit for exams organized by the Imperial Board of Astronomy. They would be selected as interns once they passed the exam, but would be sent back to where they came from if they failed and their recommenders would be punished.

Local Yin and Yang officials provided guidance to the interns and helped them manage Qiaolou (local timing system), administer altars, pray for rain, and conduct Jiuhu, a kind of ceremony that took place when solar or lunar eclipses occurred. The fact that the exam was managed by the Imperial Board of Astronomy was a strong indicator of how firmly the Imperial *Tianxue* Institute tightened its grip on Yin and Yang.

6. Examinations conducted and the process of recruitment of personnel for imperial *tianxue* agencies

An outlaw surreptitiously studying *Tianxue* in a former dynasty was oftentimes well received in a new dynasty as a credited official responsible for the Imperial *Tianxue* Institute. Later, the institute took charge of recruitment, mainly targeting talents from across the society and later providing training. Green hands, dubbed Tianwen Sheng (students studying astronomy), were selected locally.

Locally recruited *Tianxue* interns would come across thorny problems (or what now is known as "politically sensitive"). As was mentioned

earlier, *Tianxue* was banned from private practices. So it is quite para-
doxical that, on the one hand, a good citizen would never practice it while,
on the other, *Tianxue* was the soul for the Yin and Yang system and was
therefore inevitable for officials and interns to touch upon some aspects
and learn about it. As for how to provide an outline for the exam, *Volume
Seven of Sitianjian of Mishujian Zhi (Historical Documents of the Yuan
Mishujian)* gave an outline for the exam content.

*The recruitment exam, composed of six questions, was held every
three years. Anyone that passed the exam would become Sitiansheng, and
have access to further studies by reading Wuke Jingshu. After learning,
they would sit another exam and once they passed it, they would become
official members.*

It is a little difficult for modern readers to understand the perplexing
ancient language. To better understand it, we need to know Caozeren as
non-governmental Shushi (Yin and Yang practitioners). If they passed the
exam that was held every three years, they would become Sitiansheng,
students to Sitianjian, imperial astronomers. During their stints there, they
could study Wuke Jingshu, banned from private practice, before taking
another exam. Those who passed the exam would be officially appointed
as Changxing Renyuan, official members of the Imperial *Tianxue*
Institute.

What is more fascinating is that model questions and textbooks were
specified in Volume Seven of Mishujian Zhi. Caozeren were allowed to
learn from the following textbooks privately:

1. *Xuanming Li (Xuanming Calendar)*
2. *Futian Li (Futian Calendar)*
3. *Dili Xinshu (New Boo of Geography)* by Wangpu
4. *Hun Shu (Marriage Contract)* by Lu Cai
5. *Zhouyi Shifa (Yin & Yang Divination Methods)*
6. *Wuxing (Five Stars)*

There were a total of 10 questions of 4 types.

1. Questions concerning Lifa (Calendar)
   How to calculate Heng Qi Jing Shuo, some elements according to
   Xuanming Li?

How to calculate the exact position of the sun according to *Futian Li*?

2. Questions concerning *Hun Shu (Marriage Contract)*
    How many days are there in the first lunar month when Yin and Yang do not meet each other?

3. Questions concerning *Dili Xinshu (New Book of Geography)*
    If the Bagua position has got qi that connects it with that of jiuxing, can you tell whether it is beneficial or harmful to the capital city?
    Which acupuncture point is Wuxingqinjiaomingde?
    Which specific time should a dead person, surnamed Shang and born in September, Year Dingmao or 1027, be buried?

4. Questions concerning divination
    How to apply Yishi Shu to determine whether a person can seek wealth successfully on Bingchen, May, Year Dingchou?
    How to apply Liurenshu, a method of divination, to the way things work from 3 a.m. to 5 a.m., Jiazi, January?
    How to apply Sanming Shu (fortune-telling approaches) to calculate whether a person, born in Maoshi, during a period from 5 a.m. to 7 a.m., 20th May, Year Yichou, can land a decent government job?
    Give specific explanations of Qiqiang Wuruo (7 strong and 5 weak zodiacs) by applying Wuxingshu (5 stars analysis strategy).

Each time the exam was organized, 6 out of the above 10 questions would be selected. These questions are indicative of where the boundary between illegal practicing of Tianwen and legitimate practicing of Yin and Yang lies. They also show what Yin and Yang Shushi were dedicated to the general public, namely, sorting out calendars, telling fortunes, and picking auspicious days for any important events.

By the way, if you are interested in examination studies, you could turn to these textbooks and test yourself with these questions as the source.

7. Zhang Zixin's private *tianxue* activities: The only exception in Chinese *tianxue* history and a myth?

Although the *Tianxue* Institute had been strictly run by the government, there were cases of people practicing the system in private. A case in point is Tianxue-related activities carried out by Zhang Zixin in the 6th century. As shown in the historical documents, Zhang's Tianxue-related activities were of a private nature. If that is the case, it would be the only exception before the end of the Ming Dynasty.

In his books, Volume 49 of *Beiqi Shu* and Volume 89 of *Bei Shi*, Zhang Zixin made brief records without mentioning any of his *Tianxue* activities but for a story of predicting the future through Fengjiao. This is comparable to Zhuge Liang, a person of wisdom and resourcefulness, in the *Romance of the Three Kingdoms*, who was described as too resourceful to be human.[15]

Zhang Zixin, who was from Qinghe, Hebei, was a quiet learner of literature and medicine. Being a hermit at Bailu Mountain of Heilongjiang Province, Zhang occasionally paid visits to Beijing. He was well received by Wei Shou and Cui Jishu. Later, State Wei tried to recruit him as Dafu, or advisor, but failed to do so because he was not in Ye. After State Qi collapsed, he died.

Zhang's Tianxue-related activities were recorded in *Tian Wen Zhi of Sui Shu (The History of the Sui Dynasty: Astronomical Treatise)*.

In the biographies of technical figures from different dynasties, such stories of foretelling good or ill luck are too numerous to mention.

The book says: "At the end of Wei, Zhang Zixin, who's from Qinghe County, was a master of Lishu. But he became a hermit living in an ocean island, observing stars and moons with an armillary sphere for over 30 years. He finally had gained an initial understanding of the movement of stars and moons.[16]

The record shows that Zhang Zixin carried out individual Tianxue activities while he lived in seclusion on an island to evade wars. What he did had more to do with a personal hobby than government-related goals.

When the Ge Rong Riot broke out from 526 to 528, Zhang Zixin started his Tianxue-related activities on an island, and they lasted roughly 30 years until 557. A series of political coups happened after that, and

---

[15] Stories about predictions of good or ill luck abound when it comes to biographies on astrological and divination know-hows in historical records of varied dynasties.

[16] The two records possibly suggest that Zhang Zixin is a name shared by two different people as they had different native places and lines of profession. But as no solid evidence can be provided, we can do no more than cast a doubt.

Zhang Zixin could not name which dynasty he was in because he was literally a hermit. According to Beiqi Shu, he assumed offices in Northern Wei Dynasty and had contacts with Weishou, who was from Beiqi Dynasty and the author of Wei Shu. If Zhang Zixin were born after 557 when Beiqi (Northern Qi Dynasty) collapsed, he should have probably lived for another 20 years after his 3-decade stay in the island.

What Zhang achieved in the island was crucial to Tianxue's development. In the history of astronomical studies, what has long been proudly spoken of is the astronomical observances made by Tycho, a Danish astronomer, who gave merely 22 years to the cause.[17] In comparison, Zhang Zixin was more dedicated.

According to *Tian Wen Zhi of Sui Shu (The History of the Sui Dynasty: Astronomical Treatise)*, armillary sphere, a huge instrument that was very often used for Tianxue studies, was used by Zhang Zixin while on the island. The use of this device, which had to be approved by the emperor for assembly and use, was indicative of the fact that what Zhang did must have been given the green signal by the king himself. As such, since the Han Dynasty, each such device assembly was recorded.[18] And no one dared to build one privately.[19] Would Zhang Zixin be an exception? What made him an exception? These are questions that merit pondering.

He could not have been alone when observing and studying Tianxue for such a long time in the island. He needed provisions and care. And assistants were needed for device operation, record-keeping, and calculations. It is possible that he was part of a group dedicated to Tianxue studies and he was plausibly the leader of that group for 3 decades. What made Zhang Zixin so emboldened to practice Tianxue amid such a tough ban on any individual involved in the discipline? It is hardly convincing when one says that it was relaxed control over the matter during war times that made him so fearless.

---

[17] Please refer to Jiang Xiaoyuan. *Biography of Tycho, Scientists Section of Biography on World Renowned Scientists*. Science Press, (1990), pp. 8–34.

[18] Further discussion will be made on Tianxue instruments in following chapters.

[19] All devices and books involving Tianxue are banned from personal collections, otherwise any violator will be dealt with capital punishment. Still less large Tianxue devices like Hunyi. Please refer to *Tianxue Zhenyuan (The Truth of the Sciences of the Heaven)*. pp. 62–65.

What was the major progress Zhang made after going though such strenuous efforts? It was the discovery of non-uniformity between the apparent motion of the sun and planetary motion, a fact known to Greek astronomers for a long time, yet remained unknown to the Chinese until the 6th century. Zhang's discovery is believed to be a strong boost to the development of Lifa (calendar) in the Sui and the Tang *dynasties*.[20]

That said, my studies a couple of years back revealed that calendars of the Sui and Tang dynasties, which were made possible by Zhang Zixin's discovery, had a close bearing on the Babylonian mathematical astronomy of the Seleucus Age.[21] This indirectly calls into question what Zhang Zixin had discovered. Joseph Needham also assumed that Zhang probably drew inspirations from Indo-Greek astronomy.[22] And the assumption is not unprompted. We can find some hints by comparing records from those two periods.

The five planets (in the solar system) have their own preferences in their course of travel. They stay for a longer time and travel at a slower pace if they find and meet stars to their liking, and vice versa. — From *Sui Shu-Tian Wen Zhi* (2), describing one of Zhang's discoveries.

When two plants, or even three are close to each other, they develop likes and dislikes for each other. The Indian Calendar is based on the so-called Jiuzhi temperament (Nine Stars), which each temperament taken into consideration. They would move fast if they liked each other and move slowly while departing from each other. — From *Xin Tang Shu-Li Zhi Volume 3*.

Using likes and dislikes, or temperament, to explain some phenomena in planetary motion is indeed found in Indian astronomy and astrology. But this is also clearly attributed to Zhang Zixin, which at least suggests that there is a close relationship between the two.

Zhang Zixin is one of the most mysterious myths in the Chinese history of astronomy.

---

[20] Please refer to *History of Astronomy in China*, 29, p. 156.

[21] Please refer to the following two passages of mine: 1, the Relations between Babylon and China in Astronomy. Acta Astronomica Sinica, (1988), Vol. 29. No. 3 and 2, Star Movement Theory between Babylon and China. Acta Astronomica Sinica, Vol. 21. No. 4. (1990).

[22] Joseph Needham. *Science and Civilization in China*, Vol. 4. p. 531.

Chapter 5

Astronomical Phenomena and *Tianxue* Literature (1)

1. Astronomy observatory logs

As ancient Chinese believed that "Heaven hangs out its (brilliant) figures from which are seen good fortune and bad", astronomical phenomena were seen as revelations of the divine, symbolizing Providence's commending or criticism of worldly emperors' deeds or wrong-doings, which were also seen as prophesies and warnings about good or ill luck in the world. So, it was natural that various astronomical phenomena had been carefully observed and recorded, using which astronomers could sense the Providence's will and heart. The only way of getting to know what an astronomical phenomenon meant was to carefully observe and record it.

Chapter 4 mentioned the various official positions recorded in Zhou Li-Chun Guan Zong Bo (*Spring Official Zongbo of the Rites of Zhou*). Among them, *zhanmeng* was in charge of season, observing the earth and the sky, and telling *yin* and *yang*; *baozhangshi* was in charge of changes of stars and observed the celestial changes to fortell good or ill luck based on the movement of the sun and the moon, or trying to foretell the changes in the territory by observing the changes of stars. Such contents as "observing the good and evil spirits in the world from the

12-year-cyle perspective, telling the omens of the good and the evil, such as floods, droughts, harvests and famines through observing clouds and telling the harmony between the earth and the sky as well as the special evil and auspicious omens" were what an observatory (Lingtai) was all about. Theoretically, a Lingtai had professionals observing clockwise astronomical phenomena and floating clouds, with results entered in observatory logs.

Unfortunately, astronomy observatory logs in real objects have seldom been preserved to the present day.[1] Luckily, Fang Hao, a church scholar, accidentally found a paper parcel while he was studying at the Beitang Library of then Beiping (Beijing) in 1946. The parcel contained incomplete parts of ancient scripts and four sheets. The four sheets are the logs of the observatory of Qin Tian Jian (The Imperial Board of Astronomy) in the Jiaqing Period (1796–1820) of the Qing Dynasty, which is the old observatory at Jianguomen of Beijing. Though not that old, the logs, with their ink markings, are most precious. The following are Sheet 2 and Sheet 3.[2]

Sheet 2
    Day 15 of Slight Cold, Ding Mao (the fourth of the cycle of the sixty in the Chinese calendar), January 20, 1815 (19th Year 19 of Jiaqing Period)
    Logs of Observatory, Official on duty, Wuguanjianhou, records 9 times, Lu Peng (Signature)
                    Boshi records 5 times, Chang Xing (Signature)
                        7:28–14:34, Shift 1, Astronomy Student Li
                                    Weisong (Signature)
                        16:47 p.m.–9:26 a.m., Astronomy Student,
                    Zhang Pengling (Signature), 3–5 a.m., Shift 3
        3:00–9:00 a.m., Huang Dequan, Wang Guangyu

---

[1] But it is still possible for the real objects of the logs to be found. It is said someone once saw rather complete observatory logs in Archives of the Imperial Palace.

[2] Quoted from Fang Hao, *History of Exchanges between China and the West*, Yuelu Publishing House, (1987), pp. 725–728. The original format of the sheet is kept, except being changed into horizontal format.

5:00–9:00 a.m., Northwest wind, Overcast with slight sunshine 9:00–11:00 Northwest wind, Overcast with slight sunshine, 11:00 a.m.–13:00 p.m., Overcast with slight sunshine

1:00–3:00 p.m., Overcast with slight sunshine, 3:00–5:00 p.m., Overcast with slight sunshine 5:00–7:00 p.m. 7:00–9:00 p.m. 9:00 a.m.-13:00 p.m., Yu Zhongji, Huang Defu 1:00 p.m.–3:00 p.m., Bao Quan 3:00-9:00pm, Sun Qiyuan, Si Zhaonian

At 17:00–19:00 Northwest wind, Overcast with a little moonshine, 17:00–19:00, Li Weisong
19:00–21:00, Northwest wind, Overcast with a little moonshine, 19:00–21:00 Wang Guangyu

21:00–23:00 Northwest wind, Overcast with slight moonshine and starlight   21:00–23:00   Yu Zhongji

23:00–1:00 a.m. Northwest wind, Overcast with slight moonshine and starlight	23:00–1:00 a.m.		Huang Defu
1:00–3:00 Northwest wind, Overcast with a little moonshine and starlight	1:00–3:00		Huang Dequan
3:00–5:00 Northwest wind, Overcast with a little moonshine 5:00–7:00 Northwest wind, Overcast with a little moonshine and starlight	Geng 5	Xiaoke	Bao Quan Sun Qiyuan Si Zhaonian

At noontime, range quadrant employed to measure the height of the sun, the winds and clouds, Yi Zhang Zhong Biao, north shadow of the side long and south-north round shadow longer

Hand-over and Take-over Seal of Apparatuses on December 11, Year 19 of Jiaqing Period (January 20, 1815).

Sheet 3

The Waking of Insects Day 1, Ding Si, February 7, Year 21, Jiaqing

Logs of Observatory, Official on duty, Wuguanlingtailang, records 8 times, Jin Cheng (Signature)

Boshi records 5 times, Na Min (Signature) 6:20–16:20, Shift 1, Astronomy Student Bai Songxiu (Signature)

17:40–7:40, Astronomy Student Xu Zhiping (Signature), 3:00–5:00 a.m., Shift 2

5:00–7:00 a.m., Northeast Breeze, Overcast with slight sunshine, 3:00–9:00 a.m., Li Jun, Sun An, 7:00–9:00 a.m., Northeast Breeze, Overcast with slight sunshine

9:00–11:00 a.m., Northeast Breeze, Overcast with slight sunshine, sun taking on halation pale yellow in Weisu

9:00 a.m.–13:00 p.m., Li Wenjie, Tian Chen, 11:00 a.m.–13:00 p.m., Northeast Breeze, Overcast with slight sunshine, sun taking on halation pale yellow in Weisu

1:00–3:00 p.m., He Yuanfu

1:00–3:00 p.m., Northeast Breeze, Overcast with slight sunshine, 3:00–5:00 p.m., Northeast Breeze, Overcast with slight sunshine

5:00–7:00 p.m., Northeast Breeze, Overcast with slight sunshine, 3:00–9:00 p.m., He Yuandu, He Shuben, 7:00–9:00 p.m.

17:00–19:00, Northeast breeze, Overcast with starlight, 17:00–19:00, Bai Songxiu

19:00–21:00, Northwest wind, Overcast with a little moonshine Wang Guangyu 19:00–21:00 Sun An

19:00–21:00, Northwest wind, Overcast with a little moonshine, Wang Guangyu, 19:00–21:00 Sun An

21:00–23:00, Northwest wind, Overcast with a little moonshine, 21:00–23:00, Li Wenjie

23:00–1:00, Northwest wind, Overcast with a little moonshine, 23:00–1:00, Tian Chen

1:00–3:00, Northwest wind, Overcast with a little moonshine, 1:00–3:00, Li Jun

3:00–5:00, Northwest wind, Overcast with a little moonshine, 3:00–5:00, He Yuanfu

5:00–7:00, Northwest wind, Overcast with a little moonshine, 5:00–7:00, He Yuandu, He Shuben

At noontime, range quadrant employed to measure the height of the sun, the winds and clouds, Yi Zhang Zhong Biao, north shadow of the side long and south-north round shadow longer

Hand-over and Take-over Seal of Apparatuses on March 5, 1816, the 21st year of Jiaqing Period.

We can see from the above two sheets that the observation and logging regulations were sound. Different people were responsible for different shifts and intervals, which had become routine by then. However, with the passage of time, there were more and more problems: the number of qualified personnel was declining and professional ethics was flagging. As a result, the routine might have been conducted in a perfunctory manner. This phenomenon had already emerged as early as in the Imperial Observatory of the Song Dynasty. Whether these sheets record the actual observations of the astronomical phenomena remains unknown.

The sheets Fang Hao found cannot tell all the contents of the Observatory Logs, which could be revealed from the following incident.

In 733, the 21st year of the Kaiyuan period, Gautama Siddh's son complained of not being able to participate in the modification of the calendar and submitted memorials to the throne together with Chen Xuanli, accusing Yixing's *Dayan Calendar* (Dayanli) of being a plagiarized version of his father's translation of *Jiuzhi Calendar* (*Jiuzhili*), claiming that the plagiarism was incomplete. The accusation was echoed by Taiziyousiyu Shuai Nangong, which is recorded in *Xin Tang Shu-Li Zhi Volume 3,* as follows:

[Emperor Xuanzong of the Tang] ordered Shiyushi Li Lin and Taishiling Huan Zhigui to check it with the Observatory Logs and they found that Dayan Calendar got 70–80% correct, Linde Calendar 30–40% correct whereas Jiuzhi Calendar only 10–20% correct. Therefore, Gautama Siddh's son's accusations against Yixing was rejected.

Emperor Xuanzong ordered the use of the sole criterion of practice to test the truth to give a verdict on the dispute. *Dayan Calendar* and

Linde Calendar are well preserved to the present day. Like other traditional calendars of China, both are based on the calculations of the revolutions of the seven celestial bodies, namely, the sun, the moon, Venus, Jupiter, Mercury, Mars, and Saturn. Hence, the fact that the Observatory Logs was used to verify the accuracy of calendars means that the Observatory Logs had to record the locations of the seven celestial bodies on a regular basis (not necessarily a daily basis). But the sheets that Fang Hao found had no such content. Therefore, we can safely speculate that the four sheets Fang Hao found only represent one kind of record of the Observatory Logs.

2. Astronomical phenomena identified by the astronomy observatory

As the ancient Observatory Logs are not available, how those on duty in the Observatory observed and what celestial phenomena they recorded remain unknown.

In ancient China, observatories were places for emperors to get access to the heaven. The celestial observations in the observatories were not astronomical activities in the modern sense but astrological divinations to the core. The purpose was to learn about the Providence's evaluations of the emperor's politics and predictions about the world's weal and woe. I have made this point clear in my book *Tianxue Zhenyuan*. Hence, what celestial phenomena needed recording by the observatories can be inferred from what celestial phenomena were divinized by the traditional astrological practice of China.

We can explore the celestial phenomena in traditional astrology from the astrological classics that have passed on to the present day.

It is rather strange that the most complete and well-known astrological classic in China was written by Gautama Siddha, an Indian astrologer whose family had lived in Chang'an (Xi'an) for generations, called *Kai Yuan Zhan Jing* (*Astronomical Treatise of the Kaiyuan Era*). There will be a special section on this classic in the later parts of this book. To avoid wandering off course, I would like to summarize the celestial phenomena in traditional Chinese astrology.

There were mainly seven categories of celestial phenomena, which were as follows:

The first category was the solar category — Using solar eclipse constellations to divine solar surface conditions. When the sun moves to each of the 28 constellations, the solar eclipse that occurs accordingly has different astrological meanings. The solar surface may take on different conditions, such as bright, color-changing, lightless, mussy clouds, of teeth and paw patterns, thorn, halo, crest, earring, wearing, hugging, bearing, straightness, overlapping, lifting, girding, and holding. All of these are special terms the ancients would use to describe the solar surface conditions. Besides, there were some imaginary or illusory conditions that could never happen. There were altogether more than 50 terms used to describe solar surface conditions.

The second, lunar category

Using lunar eclipse constellations to divine solar surface conditions, similar to solar eclipse constellation divination — Moon eclipsing five stars, which refers to the phenomenon that a lunar eclipse occurs exactly at the time that the moon and one of the five stars are in the same constellation. Different star means something different astrologically. It does not mean the moon covering one of the five stars.

Lunar movement (speed and latitude change) and lunar surface conditions, which include bright, color-changing, lightless, mussy clouds, of teeth and paw patterns, horn, ray, thorn, halo, crest, earring, wearing, hugging, bearing, seen at daytime, not full at full time, no new moon at the right time, as well as some imaginary or illusory conditions, totaling dozens, were also terms used by the ancients in this regard.

The moon moving towards constellations (When the moon moves close to or covers any of the 28 constellations, it has a different astrological meaning); the moon moving close to asterisms other than the 28 constellations (When the moon moves close to or covers any asterism other than the 28 constellations, it has a different astrological meaning); lunar halo covering constellations and asterisms (similar to the above two, but the moon has a halo, which also has various astrological meanings) — these are all some of the ways used to divine events with the help of the moon.

The third category involved planets — The brightness, color, size, and shape of various planets and their movement when they pass or move close to constellations and asterisms (such as shun, liu, ni, fu, meaning clockwise, halting to change from anti-clockwise movement to clockwise movement, anti-clockwise movement, halting to change from slow clockwise movement to anti-clockwise movement, respectively) were used for study.

The fourth category was the study of stars — The brightness and color of stars was the main aspect here.

Emergence of guest stars (nova or supernova outburst). Sometimes, other phenomena like comets were mistaken as guest star.

The fifth category included comets and shooting stars and meteors — Study of their color and shape; the behavior of comets when they approach the sun, the moon, constellations, and asterisms; shooting stars; and meteors were all factors that were studied in this regard.

The sixth category was that of auspicious stars and evil stars — There are six kinds of auspicious stars but the celestial phenomena related to them remain unknown.

There are more than 80 kinds of evil stars, but the celestial phenomena related to them remain unknown.

The seventh category was the study of atmospheric phenomena — qi (air, which is illusionary. In most occasions, it actually refers to atmospheric ray phenomena), rainbow, wind and thunder, fog, haze, frost, snow, hail, graupel, and dew were the categories assessed.

Observatories did not have to note down all the seven categories of celestial phenomena. But we have enough evidence to judge that ancient astrologists observed and recorded most of them. Evidences for this can be seen in the *Lidai Tianxiang Jilu Zongji* (*Sylloge of Records of Celestial Phenomena in Different Dynasties*).

The *Sylloge of Records of Celestial Phenomena in Different Dynasties* collected all records of celestial phenomena dated up to 1911 in *Twenty-Four Histories, Historical Records of the Qing Dynasty, Memoir of the Ming, Memoir of the Qing, Ten Historical Books*,[3] local chronicles across

---

[3] Historians categorize the following ten historical books into Ten Historical Books, namely, *Tong Dian, Xu Tong Dian, Qingchao Tongdian, Tong Zhi, Xu Tongzhi, Qingchao*

the country, and other ancient literature. The celestial phenomena include those listed in the below table.

Item	No. of Cases
Macula	300+
Aurora	300+
Meteorolite	300+
Solar eclipse	1600+
Lunar eclipse	1100+
Moon sheltering planets	200+
Nova/supernova	100+
Comet	1000+
Shooting star	4900+

Besides, in the appendixes, there are more than 200 cases recorded (such as abnormal twilight, discoloring of the sun and the moon, rainy grey, and rainy sunspot). That such a great number of records of celestial phenomena are passed down to the present day indicates that ancient Tianxue experts had been obsrving these phenomena for a long time.

3. The use of astronomical phenomena: To know providence and indoctrinate the mortals?

The compiling and publishing of *Sylloge of Records of Celestial Phenomena in Different Dynasties*[4] intends to facilitate modern scientific research with the help of ancient records. As we have often said: "make the past serve the present". However, the intention went astray from that of the ancients. Some of the treatises were prone to deify ancients, many

---

Tongzhi, Wenxian Tongkao, Xu Wenxian Tongkao, Qingchao Wenxian Tongkao, Qingchao Xu Wenxian Tongkao.

[4] This sylloge was compiled after a host of Chinese scientists worked together for three years (1975–1977) and consulted more than 150 thousand volumes. In 1988, it was finally published by Jiangsu Science and Technology Publishing House.

had modern scientific concepts imposed on them. Even if it was done out of goodwill, the act cannot be regarded as reasonable.

So, what is the intention of our ancestors in observing and recording astronomical phenomena? It was merely to know the providence and indoctrinate the mortals.

The ancient Chinese firmly upheld the belief that interaction exists between heaven and man, and astronomical phenomena foretell what is to come while praising or denunciating what has happened, which constitutes the rationale of astrology. In this regard, ancient Western civilization is in accord with the oriental one.

The development of astrology is a long-term process of accumulation. Wu Hsi (shamans, psychics who can tell the divine will) in ancient China — later evolving into full-time astronomers — observed and recorded many astronomical phenomena. By matching the records with abundant historical events, they tried to decipher the law within, which, of course, based on modern science, we will never subscribe to. However, the ancients had unwavering faith in it. For example, the phenomenon of Wuxing Jushe (five planets, i.e., Venus, Jupiter, Mercury, Mars, and Saturn converging in one of the 28 lunar mansions) reputedly arose when King Wu of Zhou attacked King Zhou of Shang.[5] Since the assault led to the most-renowned substitution of regimes, this astronomical phenomenon was believed to foreshadow a change of dynasties. This kind of logic can be applied to other phenomena as well.

When enough "laws" were accumulated, the theoretical framework of astrology came into being, which, in written form, turned into immortal literature on astrology. Theoretically, later generations could make out the divine will according to such works.

As for the indoctrination of the mortals, a similar approach — compiling of the so-called Shichuan Shiyan — was resorted to, by which annual records of astronomical phenomena and military-political affairs in previous dynasties were put in synergy with astrology theories, enabling readers to form one-to-one correspondence between the phenomena and historical events. This practice can be traced back to *Tianguan Shu* (*Book*

---

[5]The author is responsible for an important chapter in *Research Project of the History in the Xia, Shang and Zhou Dynasties*, i.e., *Astronomical Phenomena during King Wu's Crusading Against the Tyrant Zhou*. According to the result of our research, the legend is unreliable.

of Tianxue Officials) in *Shih Chi* (*Records of the Grand Historian*), as represented in the following instances.

During the reign of Qin Shi Huang, comets appeared four times in 15 years, with the most continuous one lasting for 80 days. It stretched from one end of the sky to the other. Later, troops of Qin annihilated the other six vassal states and united the nation, wiping out surrounding ethnic groups and causing innumerable deaths.

When Xiang Yu sent reinforcements to attack Qin troops in Julu, Wangshi Star slid across the sky from the west. Later, troops of other vassal states joined to massacre subjects of Qin in the west and ravage the capital Xianyang.

At the peak of the Han Dynasty (202 B.C.–220 A.D.), five stars gathered around Dongjing, one of the 28 lunar mansions. When Han troops were besieged in Pingcheng, seven loops of lunar haloes occurred around Shen and Bi (two lunar mansions). And when the family of Lü interfered with state affairs and stirred riots, a solar eclipse arose, which led to complete darkness during the day.

When insurgency raged in seven states including Wu and Chu, comets stretched for several zhangs (zhang, a unit equaling to 3 1/3 metres) and Tiangou Star (Heavenly Hound) passed across the sky on the outskirts of Liang. After troops were sent to crush the riots, Liang was flooded with the blood of the dead.

The function of *Shichuan Shiyan* was that in the name of Providence, people were inculcated with an idea that nobody escapes the divine will, which rewards the virtuous while punishing the evil in the very end.

Therefore, the reason why ancients observed and recorded astronomical phenomena laboriously can be thought of as being for the following reasons.

First, to know the Providence through composing works on astrology; and second, to indoctrinate the mortals through compiling Shichuan Shiyan.

In the following, a section will be devoted to introducing Shichuan Shiyan, and several sections will be devoted to literature on astrology.

4. Shichuan Shiyan

Shichuan Shiyan is, on the one hand, specific testimony to the belief that astronomical phenomena forecasts good and ill luck. On the other, it is a vivid textbook of political-related moral education. Hence, it assumes an important place in *Tianwen Zhi (Astronomical Treatise)* and *Wuxing Zhi (Records of the Five Elements)* of official historical records in all dynasties.

In *Tianwen Zhi (Records of Astronomy)* of *Han Shu (Book of the Han Dynasty)*, Shichuan Shiyan accounts for three-tenths of all the contents in a fixed format starting with specific dates and the description of astronomical phenomena, followed by "the diviner said", which lists out interpretations or predictions based on astrology. Finally, historical events at that time (or around that time) are listed to attest to the prophecy of astronomical phenomena. The following are two examples:

> *In April, the third year of Jianyuan Period, comets were seen to move from Tianji Star to Zhinv Star. The diviner said, "Zhinv Star forebodes change of the empress while Tianji Star forebodes upcoming earthquake". In the next October, earthquake broke out and later, Empress Chen was dethroned.*
>
> *In the sixth year of Jianyuan Period (135 B.C.), Mars lingered around Yugui (one of the 28 lunar mansions, now a part of Cancer).The diviner said, "the phenomenon forebodes fire and bereavement". In the same year, the cemetery of Gaozu (the first emperor of the Han Dynasty) witnessed fire and Empress Dowager Dou died.*

During later times, Shichuan Shiyan followed the pattern set in *Han Shu*. The enthusiasm towards Shichuan Shiyan reached its peak in *Tianwen Zhi (Astronomical Treatise)* of *Houhan Shu (Book of the Later Han Dynasty)*, which, in three volumes, excluded any other contents except Shichuan Shiyan.

From 6 A.D. (the first year of Wang Mang Regime) to 220 A.D. (the 25th year of Jian'an Period), there was a fallacy claiming that "*every change of stars indicates a warning from the Providence and foretells a change in royal affairs*". This extreme view led to an exclusive account of Shichuan Shiyan in *Tianxiang Zhi (Records of Astronomical Phenomena)* of *Wei Shu (Book of the Wei Dynasty)*. Later, considerable Shichuan Shiyan could be found in *Tianwen Zhi (Astronomical Treatise)* of *Jin Shu (Book of the Jin Dynasty)* and *Sui Shu (Book of the Sui Dynasty)*. Later generations followed this practice in historical records (except some rare cases), which developed into a tradition.

In ancient times, "Shichuan Shiyan" was a far-reaching concept. Apart from the aforementioned professional literature, it made its presence felt in numerous historical records related to astrology. In Chinese history, relatively noted astrological prophecies were nearly all coupled with records that declared their infallibility. Thereupon, here is a question — Can astronomical phenomena really forebode what's to happen while astrologists make right predictions?

From the point of modern science, the answer is definitely no. However, in many historically renowned Shichuan Shiyan, astronomical phenomena and matching events did occur during history. Can modern science explain this? Of course it can, and the reason for this is rather simple.

The key point is, as we have seen in previous sections, that given so many astronomical phenomena and more than one interpretation for each phenomenon, there can be numerous astrological prophecies. Besides, the number of historical events is fairly large — because according to an unwritten consensus, events happening three years before or after a specific astronomical phenomenon can be selected for evidence.[6] Because of that, the compiler of Shichuan Shiyan only needs to, by his own will, simply choose from all the astronomical phenomena and historical events and match them up.

Are there any false astrological prophecies? Of course, and plenty of them. But so long as they are ruled out in historical records, later generations will never know or even consider them. Similarly, is there any historical event that runs counter to a prophecy? Of course, and there are plenty of these as well. But so long as they are eliminated from Shichuan Shiyan, what readers see are solely magical testified prophecies.

Shichuan Shiyan inculcates in readers the idea that the divine will is decipherable and inescapable. But compilers know the truth. In ancient China, there was a long-standing tradition of saints imparting teachings from the Providence, the essence behind this being that informative or

---

[6] So far, consensus hasn't been found in the ancient literature on astrology, which, however, is not difficult to deduce from abundant cases of Shichuan Shiyan.

knowledgeable authorities managed to get or even force powerless ordinary people to believe their packs of lies. Shichuan Shiyan in astrology is one of the most classic examples in that respect.

Also, the "wisdom" in compiling Shichuan Shiyan is still cherished by many who live on cheating the public. For example, we often see some swindlers touting superhuman power. In their books are countless letters from readers, all of which laud the greatness of some specific maniacal practices. Putting aside the authenticity of these letters, compilers need to rule out any letters that express disappointment, doubts, or accusation and only list those filled with praises. What's worse is that they, for a long time, kept important but unfavorable facts from the masses, only publicizing those that were advantageous to them.

5. Content, value, and stories of *Kai Yuan Zhan Jing* (Kaiyuan Treatise on Astrology)

Though literature on Shichuan Shiyan is plentiful, works with just Shichuan Shiyan as the content have not been seen. However, to know the Providence, the ancients always compiled books incorporating exclusive astrological secrets, some of which are kept secret even today. In this section, we are going to introduce one of the most significant pieces — *Kai Yuan Zhan Jing* (*Astronomical Treatise of the Kaiyuan Era*).

In modern and contemporary times, unearthing ancient books is not uncommon; some examples of these are the oracle bone inscriptions from Yin ruins, volumes in Dunhuang, *Ri Shu* (*Daybook*) written on bamboo slips from the Qin Dynasty, silk manuscripts of the Han Dynasty in Mawangdui Tomb, and bamboo slips of the Western Han Dynasty from Zhangjiashan. In ancient times, despite the absence of modern archaeological excavations, books from previous dynasties were accidentally discovered at times, among which were some distinguished ones like *Shang Shu* (*Book of Documents*) on the walls of a former residence of Confucius and *Zhushu Jinian* (*Bamboo Annals*). *Kai Yuan Zhan Jing*, the most significant and unique book in the Chinese history of astrology, can also be identified as one of them.

Comprised of 120 volumes, the extant *Kai Yuan Zhan Jing* was authored by Gautama Siddha, a tianxue expert from India who served in the imperial astronomical institute of Tang. He came from a family that had left India and had long settled in Chang'an. They were very immersed in China and its cultures and traditions and, for generations, the family members had taken up important posts in the imperial tianxue institute. Compilation of the book started in the Kaiyuan Period as demanded by the emperor, and presumably finished some years later, within a short time span, though the specific year was not recorded. *Jiuzhi Calendar*, an Indian calendar recorded in the book, was said to have been fully translated by Gautama Siddha by the sixth year of the Kaiyuan Period, according to the latter part of Vol. 4 of *Records of Calendars* in *Xin Tangshu* (*New Book of Tang*), which also claimed that the book ended with *Linde Li*. From *Xin Tangshu*, we know that *Kai Yuan Zhan Jing* was compiled after the translation of Jiuzhi Calendar and finished with the Linde Calendar before the 16th year of the Kaiyuan Period, that is to say, the year of completion was between 718 and 728 A.D.

After the book was finished, it only appeared one time in official historical records. Vol. 3 of *Records of Literature* in *Xin Tangshu* (*New Book of Tang*) said, "110-volume *Kai Yuan Zhan Jing* of the Tang is compiled by Gautama Siddha". Thereafter, the booked seemed to go missing and vanished from the public view. Since studies of tianxue (astrology) were exclusively controlled by the imperial court, masterpieces on astrology became forbidden and private collections were considered a felony, naturally leading to scarce circulation of these works. Until the Song and Yuan Dynasties (960–1368 A.D.), *Kai Yuan Zhan Jing* had been lost for hundreds of years, and in the Ming Dynasty (1368–1644 A.D.), its copies could not even be found in Qin Tian Jian (The Imperial Board of Astronomy).

However, an accident in 1617 A.D. (the 45th year of Wanli Period) brought the book back to light. Cheng Mingshan, a scholar passionate about astronomical phenomena and Buddhism, happened to find an ancient book during his philanthropic act to gild a dilapidated Buddha, which turned out to be the long-forgotten *Kai Yuan Zhan Jing*. Cheng Mingshan and his brother, Cheng Mingzhe, were overjoyed by their finding of the rare book, attributing the luck to book collection and goodwill practices in their daily life.

In the preface of *Kai Yuan Zhan Jing*, the two brothers related the story of its discovery. Being the only record to account for the book's origin, it is not supposed to be given full credit. However, considering that no traces of forgery can be discerned (except for some additional materials added during the circulation), later-generation academics generally believe their story.

The brothers are even luckier given the time of discovery. In ancient China, private study of tianxue had always been strictly banned. Even in the early Ming Dynasty, the ban was still in force, as can be seen from Vol. 20 of *Wanli Yehuo Bian* (*An Unofficial History of Wanli Period*), which stated that people studying calendars would be exiled while those making calendars would be sentenced to death. It was not until the mid-Ming period that the tradition began to alter, as a saying goes, "Emperor Hongzhi loosened the ban". It was only after this that officials outside of Qin Tian Jian dared to openly discuss tianxue. Had the brothers found the book 150 years earlier, it would have been a hot potato instead of a blessing for them. They may have ended up submitting it to the local government or furtively hoarding it and at peril of committing a felony. Fortunately, public discussion of tianxue — the study of astrology — was no longer a risk in the Wanli Period.

The rare book naturally enjoyed wide circulation owing to the increased open-mindedness on tianxue. There have been many copies, three of which are now preserved in the National Library of China, different from each other in format and diction. One of the popular editions is the Hengdetang block-printed edition from the Daoguang era (1782–1850 A.D.) in the Qing Dynasty. The book is also included in *Si Ku Quan Shu* (Complete Library in the Four Branches of Literature), which gives credit to the brothers' story in its synopsis. In Mainland China, the most accessible copies of *Kai Yuan Zhan Jing* are photo-offset editions extracted from *Si Ku Quan Shu*, like those published by China Bookstore Publishing House or Shanghai Classics Publishing House, which follow the example set by the Taiwan Wenyuange edition. In view of various printed editions laden with punctuation errors and typos, photocopies are more reliable for research.

Since its reappearance, *Kai Yuan Zhan Jing* has been widely treasured for its intrinsic value. It is no exaggeration to say that the book greatly benefits researchers on the subject of history of Chinese culture, science, philosophy, and China–India cultural exchange as well as scholars dedicated to collation of before-Tang-Dynasty classics. In brief, its academic value can be concluded from following aspects.

First, the book absorbs merits from preceding astrological books to become the most important and detailed database of ancient astrology in China. As in ancient China, astrology was monopolized by the imperial court which prohibited its circulation among the masses, the literature on astrology had little chance of passing down. Actually, we can hardly find complete astrological works aside from some related fragments in *Tianwen Zhi (Astronomical Treatise)* of specific official history books. *Lingtai Miyuan (The Secret Garden of the Observatory)*, compiled by Yu Jicaiin the Northern Zhou Dynasty (557–581 A.D.), has only 15 volumes (those reedited by Wang Anli in the Northern Song Dynasty) left now, as we are yet unable to catch a glimpse of the original copy. The 10-volume *Yisi Zhan*, authored by Li Chunfeng in the Tang Dynasty, is a relatively elaborate one. Other astrological works or copies are scattered in places like Dunhuang or Japan. A few Ming Dynasty pieces that Joseph Needham mentioned several times were not in the mainstream as orthodox astrological works. Compared with the aforementioned works, *Kai Yuan Zhan Jing*, with 120 volumes, is definitely a glorious masterpiece with all-inclusive content and complete structure. It incorporates the following:

one volume of cosmology
one volume of astronomy
one volume of astrological rules
one volume of Linde Calendar
one volume of Jiuzhi Calendar from India
one volume of Jinian and Zhanglü (important parameters of ancient calendars)
five volumes of star maps (locations of stars)
one volume of the heaven
one volume of the earth
six volumes of the sun
seven volumes of the moon
forty-two volumes of five planets
four volumes of twenty-eight lunar mansions
four volumes of Shi's divination

two volumes of Gan and Wuxian's divination
five volumes of meteors
one volume of parasitic stars
eight volumes of guest stars
three volumes of inauspicious stars
three volumes of comets
twelve volumes of weather types
ten volumes of vegetation, mansions, wares, and people and beasts

The book represents the culmination of ancient Chinese astrology. It extensively cites obscure masterpieces before the Tang Dynasty, such as *Huangdi Zhan*, *Haizhong Zhan* and *Jinzhou Zhan*, owing to which some of their contents were preserved.

Second, the book preserves ancient materials about star observation, including works of thus far the oldest three schools on astrology, which were separately founded by Mr. Shi, Mr. Gan, and Mr. Wuxian. Mr. Shi refers to Shi Shen (or Shi Shenfu) while Mr. Gan, is Gan De, both of whom were renowned astrologists in the Warring States Period (475–221 B.C.). Their masterpieces have all gone missing, except for records of star observation contained in *Kai Yuan Zhan Jing*. In the materials of Gan De, a record of the Jupiter is most astonishing.

This year, the Jupiter moves to the section of "Zi", one of the twelve earthly branches (ancient Chinese created twelve earthly branches according to the Jupiter cycle and employ them to number the years), appearing in the morning and disappearing at night along with two constellations—Xu and Wei. It looks immense with halos, surrounded by smaller red stars, the so-called companions.

According to the research of Xi Zezong, an authority on the history of astronomy, it indicates that more than 2,300 years ago, the Chinese had already observed the satellites of Jupiter with their naked eyes and had given unequivocal records, as opposed to the general belief that Galileo was the first to discover them at the beginning of the 17th century with the help of telescope.[7] *Kai Yuan Zhan Jing* also contributed to the systematic preservation of star divination by Gan, Shi, and Wuxian.

---

[7] Xi Zezong. *Gan De's Discovery of Satellites of Jupiter Two Decades Before Galileo*[J]. *Astrophysical Journal*, Vol. 1, (2) (1981).

Third, the book records all the basic data of calendars known before the 8th century. Since *Shih Chi* (*The Records of the Grand Historian*) pioneered in containing astronomical materials in *Tianguan Shu* (*Book of Tianxue Officials*) and *Li Shu* (*Book of Calendars*) and *Han Shu* (*Book of the Han Dynasty*) in *Tianwen Zhi* (*Astronomical Treatise*) and *Lüli Zhi* (*Records of Calendars*), the later generations followed the practice so that basic data of different calendars had written records. What makes *Kai Yuan Zhan Jing* stand out is that it covers kindred resources in the pre-Qin period and also redresses defects in official records after Qin and Han *dynasties*.

Fourth, the book incorporates the *Jiuzhi Calendar*, a Chinese version translated from the ancient Indian calendar. The so-called Jiuzhi includes the sun, the moon, five planets, and two imaginary planets Rahu and Ketu (they are in fact the ascending node and apogee on the moon's path).[8] *Jiuzhi Calendar* in *Kai Yuan Zhan Jing* only records techniques to calculate solar–lunar movement and eclipse, and assertions cannot be made of their completeness. Despite all these, the calendar has become an invaluable resource on studies of ancient astronomy in India and ancient astronomic exchange between China and India. As Indian astronomy originated from ancient Greece, we can see abundant Greek-flavored contents in *Jiuzhi Calendar*, such as ecliptic coordinates, the algorithm and table of sine function, etc. By now, the calendar has been translated into English and introduced to the West.

Fifth, the book is a collection of ancient divination books. It quotes from 82 kinds of divination books, many of which are no longer available, thus rendering it much more precious. Sun Jue, a scholar in the Ming Dynasty, compiled *Gu Wei Shu* after collecting previous divination books. However, the books quoted there hardly overlap with those in *Kai Yuan Zhan Jing* because the two works were composed in different dynasties. In the era when Sun lived, what Gautama Siddha had quoted were mostly lost.

---

[8] In previous domestic treatises on this issue, Rahu and Ketu were falsely believed to be the ascending and descending nodes on moon's path. The truth is that Rahu is the ascending node on moon's path while Ketu apogees on moon's path. Please refer to Niu Weixing. *A Probe into Rahu and Ketu's Astronomical Meaning*[J]. *Astronomical Journal*, Vol. 35, (3) (1994). See also Section 3 of Chapter 9.

The above five aspects were only briefly touched upon. Part of the materials on star observation in the book, such as the star catalogue of Shi Shen, may not have been produced by the end of the Han Dynasty (202 B.C.–220 A.D.), according to mind-boggling conclusions drawn from latest research.[9]

On the one hand, *Kai Yuan Zhan Jing* was authored by a Chinesized Indian expert on tianxue. On the other hand, with the support of the then emperor it enjoyed abundant resources from imperial collections. These advantages render it incomparable, and it is endowed with more mystery by its reappearance after hundreds of years of obscurity.

6. Li Chunfeng's *Yisi Zhan* (*Yisi-year Divination*)

Among all the astrological masterpieces in ancient China, *Yisi Zhan* (*Yisi-year Divination*), a work by Li Chunfeng (602–670 A.D.) in the Tang Dynasty, is second only to *Kai Yuan Zhan Jing* in terms of significance.

The book is so named because the author estimates that "Shangyuan" should be Yisi year, when the sun, the moon, and five planets are lined up on the first day of the eleventh lunar month (namely the Winter Solstice).[10] Among all the works that are attributed to Li, this one is the most reliable. With respect to reputation or status in the Chinese history of astrology, Li is superior to Gautama Siddha. The popularity of *Yisi Zhan* can be attested to by such books as *Records of Literature* of *New Book of Tang*, *Zhizhai Shulu Jieti* (*Zhizhai Content Explanation*), *Wenxian Tongkao* (*Comprehensive Investigations of Records and Documents*), etc.

The main content of the ten volumes is as follows:

---

[9] A chapter in the latter part of this book is devoted to this topic.

[10] Shangyuan, in the eyes of ancients, is an ideal start to make a calendar, which, by modern standards, means the time when the sun, the moon and five planets are on the same ecliptic longitude. The interval between Shangyuan and the finishing year is called "Shangyuan Ji'nian". Owing to the difficulty in determining Shangyuan and the great influence exerted by occultism, Shangyuan can be estimated to be a primordial era which leads to incredibly long Shangyuan Ji'nian, like the one in Tang-Dynasty Dayan Calendar(90 million-plus years) and a more unbelievable one in the Jin-Dynasty revision of Daming Calendar (380 million-plus years).

Vol. 1

Introduction of astronomical data and instruments; divination on astronomical phenomena like "blood rain" and "flesh rain" (most of them are not likely to happen in real life); divination on solar phenomena like solar eclipse and floating clouds near the sun

Vol. 2

Divination on lunar eclipse and lunar aureole; divination on the moon's approach towards or overlap with constellations out of twenty-eight lunar mansions

Vol. 3

Theory of Fenye;

examples of divination;

divination related to numbering years, months, and dates;

Xiude (cultivation of virtues);

Bianhuo (answers to puzzles)(lost); work ethics of astrologists

Vol. 4

Theories and data of five planets; astrological theories of five planets; divination on five planets' approach towards or overlap with constellations out of twenty-eight lunar mansions; divination about the Jupiter

Vol. 5

Divination about the Mars; divination about the Saturn

Vol. 6

Divination about Venus; divination about Mercury

Vol. 7

Divination on meteors' approach towards or overlap with the sun, the moon, and five planets;

Divination on meteors' entrance into every lunar mansion; divination on guest stars' approach towards or overlap with constellations out of twenty-eight lunar mansions (guest stars refer to comets, novae, or supernovae)

Vol. 8

Divination about comets;

divination about parasitic and inauspicious stars;

divination on clouds' colors and locations;
divination about clouds

Vol. 9
Wangqishu (a method to tell good or ill luck)

Vol. 10
Wind divination

Overall, the content in Volumes 1 to 8 is very similar to that in *Kai Yuan Zhan Jing*. Likewise, *Yisi Zhan* also quotes from previous astrological masterpieces, many of which coincide with those in *Kai Yuan Zhan Jing* . The difference lies in that the quotations in *Yisi Zhan* are more brief and, to a great extent, dependent on the author's discretion. The era Li Chuanfeng lived in was only about a decade earlier than that of Gautama Siddha; so both, as directors of the imperial *tianxue* academy, would have had access to the same batch of astrological books in the royal library, which explains the similarity. What distinguishes Yisi Zhan is the content of Vol. 9 and Vol.10, which occupies only a small proportion in *Kai Yuan Zhan Jing*.

Vol. 9 is specially devoted to Wangqishu (a method of telling good or ill luck), which, in ancient times, was mainly discussed in the military field to predict the result of a battle or the rise of a ruler through observations of Qi. The Qi here is a rather supernatural and elusive thing. From the perspective of modern science, sometimes it is considered an atmospheric phenomenon, but most of the time, it can be hard to ascertain. Despite the elusiveness, it is by no means groundless, and we can still see the heritage of Wangqishu from expressions like "Qifen" and "Fenwei" (atmosphere).[11] There are altogether 14 entries in Vol.9 of *Yisi Zhan*:

Qi of a Sovereign; Qi of a General; Qi of Victory in War; Qi of Defeat in War; Qi of City Prosperity; Qi of City Massacre; Qi of Ambush; Qi of Brutal Troops; Qi of Battle Arrays; Qi of Conspiracy; Qi of Good or Ill Omens; Qi of Nationwide Abnormal Scenes; Clouds' Entrance into

---

[11] Even today, we often hear utterances like "Qifen (atmosphere) in the meeting place is very tense". The Qifen here is also intangible but perceptible. Maybe that's the heritage of ancient Wangqishu.

Every Lunar Mansion; Clouds' Entrance into Constellations Out of Twenty-eight Lunar Mansion

As astrologists served as advisers or counselors to the monarch, whose major concerns were political and military affairs, learning Wangqishu was indispensable for them.

Vol. 10 gives an exclusive and detailed explanation of wind divination. In the 120-volume *Kai Yuan Zhan Jing*, the topic is only mentioned in one volume; by contrast, in *Yisi Zhan*, nearly one-fifth of the content is devoted to wind divination (each of the first nine volumes contains about 10,000 words while Vol. 10 counts almost 30,000 words). There are altogether 42 entries in Vol. 10, and these are as follows:

Houfeng (wind measuring)

Zhanfeng Yuanjin (measuring how far or near the wind is)

Tui Fengsheng Wuying (identifying the five sounds in winds)

Wuyin Suozhuzhan (divination on omens of the five sounds)

Wuyin Fenzhan (divination according to the five sounds)

Lun Wuyin Liushu (testing the six properties of the five sounds)

Wuyin Shouzheng Shuori Zhan (five-sound divination on the first day of a lunar month)

Wuyin Xiangdong Fengzhan (divination of the five sounds and the wind sounds)

Wuyin Mingtiao Yi Shangzu Qi Gongzhai Zhong Zhan (divination of the wind sounds in tress in palaces and residences)

Tui Suiyue Rishi Gande Xingsha (estimating the deities on duty in different days, months, and years)

Lun Liuqing (estimating the six emotions of people)

Yin-yang Liuqing Wuyin Licheng (outline for estimating *yin* and *yang*, six emotions, and five sounds)

Liuqing Fengniao Suoqi Jiashi Zhan (divination of six emotions and variations of winds)

Bafang Baofeng Zhan (divination on storms from eight directions)

Xingdao Gongzhai Zhong Zhan (divination on travelling and palaces)

Shi'er Chengfeng Zhan (divination on winds in 12 2-hour periods)

Zhu Jiebing Feng Zhan (divination on battle fields)

Zhu Xian Cheng Feng Zhan (divination of the fall of cities)

Ru Bingying Feng (divination on the winds in military camps)

Wuyin Ke Zhu Fa (divination of five sounds on guests and hosts)

Sifang Yidi Qinjun Guofeng Zhan (divination on foreign invasion from four directions)

Zhanguan Qian Mianzui Fa (divination on officials' exemption from punishment)

Hou Zhao Shu (divination on Imperial Edicts)

Houshe Shushu (divination on remission letter)

Hou Dashe Feng (divination on remission)

Hou Dabing Jiangqi (divination on army assembly)

Hou Dabing Qie Jiesan (divination on army assembly and dismissal)

Hou Da Zai (divination on big disasters)

Hou Zhugong Guike (divination on distinguished guests of nobility)

Hou Dabing Gongcheng Bing Shengfu (divination on storming the castle and city)

Hou Zei Zhan (divination on thieves)

Hou Sang Ji (divination on death and disease)

Hou Siyi Ru Zhongguo (divination on entry of foreigners in China)

Zazhan Wanghou Gongqing Erqianshi Churu (divination on nobles' harvest)

Zhanfeng Tu (divination of winds)

Zhan Bafeng Zhi Zhuke Shengfu Fa (divination of eight winds on victory and defeat of guest and host)

Zhan Fengchu Jun Fa (divination of winds on dispatching troops)

Zhan Xuanfeng Fa (divination on whirlwinds)

Sanxing Fa (divination on three penal punishments)

Xiang Xing Fa (divination on two penal punishments)

Wu Mu Fa (divination on five tomb methods)

Desheng Fa (divination on deity of virtues)

From these entries, we can see that *Yisi Zhan* is virtually a complete guide on wind divination. In fact, this is what Li Chunfeng intended, as he wrote in Vol. 10:

I admired the art (wind divination) so much that I read innumerable books for research. Since the time of Yi Feng (an expert on classics in the

Western Han Dynasty), there have been nearly a hundred books on the topic, some detailed while some brief, some well-founded while some misleading. I remove the nonsense like bizarre, shallow statements and false methods and then compose a volume... Hopefully, these words are lucid but explanations elaborate and theories convincing, which can be turned to by future pursuers of wind divination.

It seems that Li has made painstaking efforts to study and explain wind divination, which, in ancient China, was a popular divination art carried out by observing the wind from all directions. Based on Wuxing (Five Elements) and Bagua (Eight Diagrams), the art features independent theories, a set of special terms, and expressions, using rhetoric devices such as parallelism and cohesion to support statements. But its basic concepts and theories are within the scope of astrology. Here, I will give some clear examples of divination:

If the wind blows from south on your way, you are bound to be treated with wine and delicacies. If the whirlwind sweeps from the door to the side of the principal room, the eldest son will commit theft.

If the wind whirls into the well, the wife wants to scheme with others to murder her husband. On the day of Zhugong, the strong wind comes from Jiao freezing and swift, which augurs the city besiegement by mighty troops.

If a gale destroys houses and trees at the noon, the city will fall into enemy hands within nine days.

From today's vantage point, these statements sound mostly ridiculous.

6.1. Lingtai Miyuan (*The Secret Garden of the Observatory*)

After *Kai Yuan Zhan Jing* and *Yisi Zhan* is *Lintai Miyuan* (*The Secret Garden of the Observatory*), a relatively complete work on astrology. This masterpiece was finished in the Northern Zhou Dynasty (557–581 A.D.), much earlier than the other two Tang Dynasty works. One section of the *Records of Classics, Book of the Sui Dynasty,* reads that the 115-volume book was authored by the imperial astronomer Yu Jicai. Unfortunately, its original version is lost, leaving only the extant one reedited by Wang Anli (1034–1095 A.D., younger brother of Wang Anshi, a renowned politician and litterateur) and others in the Northern Song Dynasty, which, have a total of 15 volumes, presumably an abridged edition. Vol.1 records several star maps; however, their origins elude us since the book has undergone a revision in the Northern Song Dynasty.

Finally, we have to note that two other important works on astrology are also related to Li Chuenfeng, namely,

> *Tian Wen Zhi of Jin Shu (Astronomical Treatise, Book of the Jin Dynasty)* and *Tian Wen Zhi of Sui Shu (Astronomical Treatise, Book of the Sui Dynasty)*.

That leads to another important topic — astrological materials in official history books.

7. Three records on *tianxue*

In official history books of all dynasties, records of various topics abound, three of which are related to tianxue and often superior to others, and they are what I call *Tianxue Sanzhi (Three Records on Tianxue)*.[12] They are as follows:

Tian Wen Zhi (Astronomical Treatise): It focuses on the content, including materials on star observation, records of astronomical phenomena, Shichuan Shiyan, astronomical instruments, cosmologies, and important activities of *tianxue*.

Lü Li Zhi (Records of Calendars): In Section *Li (Calendar)*, the content is centered on calendars, including the history of calendars, terms in major calendars, disputes about calendars, history of calendar-making organizations, etc. In Section Lü, theories and data of Yinlü (Melody) are recorded, which have nothing to do with tianxue.

Wu Xing Zhi (Records of the Five Elements): It exclusively records kinds of Xiangrui (auspicious omens) and Zaiyi (calamitous omens). Some of these uncommon natural phenomena come from ancient imagination or legends.

---

[12] In respect to Tianxue Sanzhi's prime importance in official history books, please refer to Chapter 3 of *Tianxue Zhenyuan (The Truth of the Sciences of the Heaven)*.

Tianxue Sanzhi has its fountainhead in *Records of the Grand Historian*, which incorporates *Book of Tianxue Officials, Book of Melody,* and *Book of Calendars* among eight different books. In *Book of the Han Dynasty, Astronomical Treatise* is found as a counterpart of *Book of Tianxue Officials*, the combination of *Book of Melody* and *Book of Calendars* makes *Records of Calendars,* while *Records of the Five Elements* has been added. In later-generation history books, the same pattern is followed, except some rare cases (whether or not records of melody and calendars are combined is contingent).

Below is a list of Tianxue Sanzhi in different history books[13]

Records of the Grand Historian	*Book of Tianxue Officials*	*Book of Calendars*		
Book of the Han Dynasty	*Records of Astronomy*	*Records of Calendars*	*Records of the Five Elements*	
Book of the Later Han Dynasty	*Records of Astronomy*	*Records of Calendars*	*Records of the Five Elements*	
Book of the Jin Dynasty	*Records of Astronomy*	*Records of Calendars*	*Records of the Five Elements*	
Book of the Song Dynasty	*Records of Astronomy*	*Records of Calendars*	*Records of the Five Elements*	*Records of Auspicious Signs*
Book of the Southern Qi Dynasty	*Records of Astronomy*		*Records of the Five Elements*	*Records of Auspicious Omens*
Book of the Wei Dynasty	*Records of Astronomical Phenomena*	*Records of Calendars*		*Records of Propitious Omens*

---

[13] These records on *tianxue* had already been included at the completion of official history books. There are also records compiled by later generations; whereas, they are not regarded as part of the official history books.

Book of the Sui Dynasty	Records of Astronomy	Records of Calendars	Records of the Five Elements
Older Book of Tang	Records of Astronomy	Records of Calendar	Records of the Five Elements
New Book of Tang	Records of Astronomy	Records of Calendar	Records of the Five Elements
Old History of the Five Dynasties	Records of Astronomy	Records of Calendar	Records of the Five Elements
New History of the Five Dynasties	Investigation of Imperial Astronomer		
History of the Song Dynasty	Records of Astronomy	Records of Calendars	Records of the Five Elements
History of the Liao Dynasty		Records of Calendar Tables	
History of the Jin Dynasty	Records of Astronomy	Records of Calendar	Records of the Five Elements
History of the Yuan Dynasty	Records of Astronomy	Records of Calendar	Records of the Five Elements
History of the Ming Dynasty	Records of Astronomy	Records of Calendar	Records of the Five Elements
Draft History of the Qing Dynasty	Records of Astronomy	Records of Constitution	Records of Disaster and Monstrosity

The list above offers us a glimpse of three records on tianxue (despite variations on how they are named) as well as some exceptions (sometimes there are records of auspicious omens) in official history books. What's worth noting here is that these records are typical materials on astrology, which, accounts for why they should be compiled by such experts as Li Chunfeng. It is easy to comprehend that *Astronomical Treatise* and *Records of the Five Elements* pertains to astrology. However, in truth, much of the content in *Records of Calendars* also serves astrology.[14]

I have mentioned several times that astrology in ancient China was a forbidden discipline and private study of *tianxue* constituted a felony. Nonetheless, official history books were open to the general public. So, here is a question — If ordinary scholars probe into the *Tianxue* Sanzhi within, does that count as private study of *tianxue*? So far, express stipulations have not been found, but according to following stories, the boundary is rather blurry.

In *Sheng'an Yishi (Anecdotes of Sheng'an)* composed by Cheng Feng, there is an anecdote of Yang Shen (1488–1559 A.D., alternatively named Sheng'an):

> *While reading Astronomy, Comprehensive Investigations of Records and Document, Emperor Wuzong of the Ming came across a star named Zhuzhang but failed to know its real identity. He asked the officials in Qin Tian Jian what on earth it was but no one could answer. So he resorted to Hanlin Academy; still, everyone was at a loss. At that time, Yang Shen stood out to give an answer. He thought it was Star Liu and supported his assumption with two records respectively in Records of the Grand Historian and Book of Han. His colleagues made a joke by saying, "though you are eloquent and knowledgeable, does it not occur to you that your answer might betray your private study of tianxue?"*

The inability of officials of Qin Tian Jian to answer the question reflects their professional incompetence, whereas those in Hanlin Academy may have concealed their answer out of fear of courting trouble, which is what can be inferred from the joke. But Yang only referred to two official

---

[14]The statement is evidenced in Chapter 4 of *Tianxue Zhenyuan*.

history books, *Shih Chi* and *Han Shu,* and that did not bring him much trouble. From the following story, we can have a clearer view of the prohibition on private study of *tianxue.*

At the beginning of the Ming Dynasty, the prohibition was followed as a heritage from past dynasties, which ruled that apart from Qin Tian Jian officials and a handful of personages, other people (including ordinary officials, soldiers, and civilians) would be accused of committing a felony if they secretly studied *tianxue.* In *An Unofficial History of Wanli Period* is an often-quoted sentence depicting the situation then — calendar learners would be exiled while calendar makers would be sentenced to death. However, since the mid-Ming period, the prohibition began to slacken, and during the Wanli Period (1573–1620 A.D.), many started to openly violate it. In 1584, Fan Shouji, an official in the Ministry of War, built an armillary sphere by himself, which was outrageous because according to the stern prohibition in previous dynasties, hiding tianxue instruments led to death, let alone privately building them. This conduct provoked such a stir that he earned flocks of spectators. So he composed *Tianguan Juzheng (Corrections by Heavenly Officials)* and defended himself in the preface by saying that:

> *Some say that the national ban on private study of tianxue should be adhered to by every man of honour. But I want to speak in defense of myself. Such a claim is, in fact, touted by those blind followers. With astrological works in official history books, people have been discussing about astrological knowledge for nearly two hundred years, which learned scholars taught and students acquired. Can the discipline be banned because of some villainous lawmakers?*

Fan Shouji argues that it is legitimate for scholars to read astrological works in official history books such as *Shih Chi* — a statement that could not have been uttered but for the loosened ban. If he had been in the early years of Ming, such remarks would have brought him much trouble, if not punishment for a crime.

---

[15] See Jiang Xiaoyuan: *Tianxue Zhenyuan (The Truth of the Sciences of the Heaven),* (2007) pp. 62–65.

Chapter 6

Astronomical Phenomena and *Tianxue* Literature (2)

Apart from *Kai Yuan Zhan Jing, Yisi Zhan, Lingtai Miyuan,* and the "Three Treatises on *Tianxue*" in the official historical literature, other remaining astrological pieces are either fragmentary, or undated, or lost abroad. A brief introduction of the outstanding ones among these will be given in what follows.

1. Shishi Xingjing (*The Star Manuals of Master Shi*) & Ganshi Xingjing (*Gan's and Shi's Star Manuals*)

Let us start with the renowned *Shishi Xingjing* and *Ganshi Xingjing*. As its name suggests, the latter is often believed to be written by Gan De and Shi Shen(fu). Gan and Shi enjoyed comparable reputations, their names frequently appearing side by side in the Han Dynasty. For example, the *Tianguan Shu of Shih Chi* recorded that "so Gan and Shi performed intimate examinations on the five planets (Jupiter, Mars, Saturn, Mercury and Venus), discovering that only Mars would retrograde". Also in the *Astronomical Treatise, the Book of Han*:

> "The ancient calendars did not reckon with the retrograde motion of any
> of the five planets, until Gan and Shi composed their star manuals,
> exposing this erratic phenomenon of Mars and Venus".

The exact time that marked China's first account of planetary retrograde is an important topic in the history of Chinese planetary astronomy. But now let us cut to the chase and focus on Gan and Shi. While people of the Han Dynasty frequently mentioned the two persons together, modern scholars tended to include their birth in the Warring States Period while speaking of the duo, though actually Gan De was probably born later than Shi according to evidence from *Ranked Biographies of Zhang Er and Chen Yu, Shih Chi* that a revered Mr. Gan had persuaded Zhang Er to forsake Chu for Han — this revered Mr. Gan is considered to be none other than Gan De. It follows that Gan lived to the Chu-Han Contention, which happened after the Warring States Period.

Yet curiously, though the post-Han historical literature frequently quoted from Gan's and Shi's works, in the calculation and divination section of the *Treatise on Literature, the Book of Han*, there is no record of their works under relevant categories, except for an entry under "divination" indicating "20 volumes of *Gan De Changliu Mengzhan* (*Gan De's Changliu Oneiromancy*)". Nevertheless, reference to their works made increasingly frequent appearance since the Eastern Han Dynasty. Some such works include the following:

> Xu Shen's *Shuowen Jiezi* (*Explaining Graphs and Analyzing Characters*) named *Ganshi Xingjing*;
> *Rhythm and the Calendar, Book of the Later Han* mentioned *Shishi Xingjing*; Liang-dynasty author Ruan Xiaoxu's bibliography *Qi Lu* said that the revered Mr. Gan wrote the eight-volume collection *Tianwen Xingzhan* (Astronomic Star Observation) and that Shi Shen wrote the eight-volume collection *Tian Wen* (Heavenly Patterns)";
> *Classics, Book of Sui* stated that "the Liang Dynasty saw the eight-volume collections of Shi and Gan", and it catalogued Shi's *Huntian Tu, Shishi Xingjing Buzan, Shishi Siqifa*, etc., which were also found in the record of *Classics, Old Book of Tang*;
> *Junzhai Dushu Zhi* (Essays on Reading the Books in the Commandery Studio) by Southern Song-dynasty author Chao Gongwu recorded "a volume of *Ganshi Xingjing*".

After a long period, it naturally followed that the Ming literature attributed the *Xing Jing* (*Star Manuals*) to Gan and Shi; however, modern scholars widely acknowledge that the manuals are works of later-generation imposters (detailed below).

Even so, modern experts in the history of astronomy, while regarding the *Star Manuals* as fakes, mostly believe that Gan and Shi's masterpieces have indeed endured — at least parts of them. Also, the astrological divinations and star catalogues of the three authorities — Gan, Shi, and Wuxian — included in the *Kai Yuan Zhan Jing* compiled by Gautama Siddha of Tang Dynasty are exactly those parts that have survived. The quotes from Shi's divinations in the book are considered by most scholars as the true bequest of *Shishi Xingjing* — actually, they often simply refer to the collection of those quotes as *Shishi Xingjing*.

The reason that *Shishi Xingjing* received serious scholarly attention (far greater than the works of Gan and Wuxian) was that besides records and divination of the twenty-eight Lunar Mansions and stars in the Zhong Guan and Wai Guan region (similar content can be found in Gan's and Wuxian's works), *Shishi Xingjing* also contained the locations of 120 stars in relation to certain constellations (ruxiudu) and the North Pole (qujidu), as well as their position on the plane of the ecliptic and the moon's path (huangdao neiwaidu), which is exactly what makes it a proper star catalogue — "Shi's star catalogue" as it is often called.

Of course, the ancient Chinese *tianxue* masters used a different celestial coordinate system from that of modern astronomy. As regards the specific meanings of *ruxiudu*, *qujidu,* and *huangdao neiwaidu*, the next chapter will give a detailed explanation. Here it is only necessary to know that the first two measurements can be converted mathematically into "declination" and "right ascension" of modern astronomy.

Theoretically, according to the methods provided by celestial mechanics, with the declination and right ascension data offered by an old star catalogue, it is possible to calculate the observation year of these data based on the axial precession theory. Hence, exploring in which age the observations of *Shishi Xingjing* were conducted became an interesting topic for many astronomical historians, especially Japanese scholars. For example, Shinzo Shinjo, Minoru Ueda, Kiyoshi Yabuuchi, and Yasukatsu Maeyama who lived in Germany have all conducted dedicated studies on it.[1] As for Chinese scholars, Pan Nai is no doubt the one who has carried

---

[1] Detailed information about these studies can be found in Pan Nai: *The History of Fixed-Star Observation in China*, Xuelin Press, (1989) edition, pp. 51–55.

out the most comprehensive research, and his conclusion was published in *The History of Fixed-Star Observation in China*:

(In *Shishi Xingjing*) the first group of stars was observed in 440 B.C. on average...The second group was observed in 160 A.D. on average... So it seems that we can arrive at such a conclusion: the star catalogue of *Shishi Xingjing* is the result of observations originally conducted in the early Warring States Period, about mid-5th century B.C., and then complemented after partial loss in about the late Eastern Han Dynasty, namely the latter half of the 2nd century A.D.[2]

Pan's conclusion has been the most widely accepted by his peers for a long time. However, recently Hu Weijia sounded a different tune, coming up with opinions quite opposite to the traditional mainstream views.

He argued that the existing Shi's Star Catalogue was in fact an astronomical achievement in the Sui and Tang Dynasties.

Drawing on textual research, Hu first pointed out that regarding Shi's Star Catalogue as a work of the Warring States Period or the Han Dynasty lacked sound evidential support from historical literature. But the past mainstream view was based on data derived from mathematical calculations with parameters like axial precession, which was rather "scientific". If there are no arithmetic results proving that Shi's Star Catalogue was indeed from the Sui and Tang *dynasties*, then the iconoclastic argument can hardly be convincing, because for many people, results of astronomical calculations are far "tougher" than textual evidence. Hu could not provide such arithmetic data, but he solved it in a stroke of genius: he applied the same method the predecessors used in calculations of the Shi's Star Catalogue to a Song Dynasty star catalogue whose observation year was already known; the result is that there was a discrepancy amounting to centuries! In this way, he, in effect, used proof by contradiction to successfully show that the modern astronomical methods our predecessors kept using on this problem might not be really applicable; consequently, their conclusions become untrustworthy. He drew the conclusion as follows:

The list of Shi's asterisms was gradually enriched; it's not until Chen Zhuo combined the observations of the three schools into a single system that it became fixed and continued to go around for centuries unchanged. Our attribution of the Chinese asterisms to Shi, Gan or Wuxian today is

---

[2] Pan Nai: *The History of Fixed-Star Observation in China*, (1989), p. 64.

exactly based on Chen Zhuo's system. Shi's Star Catalogue should have come into being after the Shi School's sky of asterisms took full shape.

The earliest twenty-eight lunar mansions catalogue and Shi's Star Catalogue we now see are from the Tang Dynasty literature; comparing these lunar mansions with reference to relevant records shows that the change in the *qujidu* of the twenty-eight Lunar Mansions is caused by new observations, not misinformation.

Notes in *Kai Yuan Zhan Jing* on the *huangdao neiwaidu* of the twenty-eight Lunar Mansions, and the *rusudu, qujidu, huangdao neiwaidu* of Shi's asterisms were seen and added after a record of the twenty-eight Lunar Mansions' *qujidu* appeared in early-Tang literature. No evidence backs an earlier observation year of these data, so they should be seen as a monumental accomplishment in early-Tang astronomy.

The popular method of using axial precession to back-calculate the observation year of star catalogues has been proved unreliable; examining the starting point of the back-calculation — the star catalogue recorded in Tang literature — also tells us that the application of such a method is unnecessary.

In this case, if the rather "scientific" conclusion is untenable, then should we return to the traditional textual research across historical literature? Nevertheless, Hu's argument offers even more edification: if his argument can hold, then it will deliver a solid punch to the widely accepted conception in science history study — that the more scientific (meaning mathematic actually) a method is, the more solid the conclusion derived. He, in fact, warned researchers in the history of science that with the limitations of the ancient literature itself, some mathematical ways that are quite trusty themselves can be infeasible or even doubtful when applied to the treatment of some ancient texts.

Of course, personally speaking, even if today's *Shishi Xingjing* were the result of observations conducted in the Sui and Tang Dynasties, that Gan and Shi have lived in the Warring States Period and have left star catalogues still deserve credence.

When the author was writing this chapter, Hu's thesis elaborating on his aforementioned conclusion was still in press.[3]

---

[3] Hu Weijia: "Study on Tang Literature-Recorded Twenty-Eight Mansions Location & Shi's Star Catalogue", published in *Studies in The History of Natural Sciences*, Vol. 17, No. 2 (1998).

2. Dunhuang manuscripts P. 2512 and S. 3326

The Dunhuang Manuscripts contained abundant historical records. Among them, P. 2512 and S. 3326 are no doubt the most important. But this section will just briefly describe them and leave them to later, more elaborate discussions.

P. 2512. The beginning part of this manuscript is lost, and the whole layout is messy; sometimes different works follow immediately after each other in a jumble, other times one single work is broken into lines and paragraphs; and the subtitles are incomplete too. But anyway, a text of about 8,500 characters is preserved, which could be ranked the most essential among all the astrological materials in Dunhuang manuscripts. Its contents can be divided into the following five parts:

1. The remaining part on astrology. Due to the loss of the beginning, the first part only contains Waiguan Zhan (Waiguan Divination), Wuxian Zhan (Wuxian Divination), Zhan Wuxingse Biandong (Divination Based on Color Changes of Five Planets), Zhan Liexiubian (Divination Based on Changes of Lunar Mansions), Wuxing Shunni (Divination Based on Planetary Progradation and Retrogradation), Fenye (Field Allocation), Shier Ci (Twelve Ci), and Jiuzhou (Nine Regions).
2. *Twenty-eight Mansions Catalogue.* It provides the names, number of stars, *judu* (distance, in degrees from one reference star of the mansion), *juxing* (reference stars) of each mansion, *qujidu* of each *juxing*, and fenye (fields) to which each mansion belongs.
3. The star manuals of Shi, Gan, and Wuxian.
4. The famous Xuan Xiang Shi.
5. A casual astrological discussion with a simple illustration of the astronomical phenomena around the sun and the moon, which does not seem like a complete work. The second, third, and fourth parts of the manuscript are of great significance and will be discussed later.

S. 3326. It is a manuscript of considerable length; its beginning part is also missing. A collection of 25 cloud pictures were preserved in the first half of the manuscript — according to the notes of the original author at the end, there should be 48 in total; under all the pictures are divination statements.

The latter half has attracted special attention from researchers in the history of science; it contains 13 star maps, the first 12 of which were drawn corresponding to the twelve *ci* (ancient Chinese division of the sky). The thirteenth delineated the "Ziweiyuan" (Purple Forbidden Enclosure), captioned "Lightening Deity" — and no obvious connection to the other 12 star maps is found.

Apparently, S. 3326 is closely related to the mainstream astrological system in ancient China: the coverage of the twelve *ci* marked in the star maps completely matches the demarcation of Chen Zhuo — more on him later — recorded in *Astronomical Treatise, the Book of Jin (1)*; and the captions of the star maps are from "Simple Examples of Fenye", the sixty-fourth volume of *Kai Yuan Zhan Jing*. These captions have been collated and corrected by Academician Xi Zezong in his research published in 1966.[4]

As to which method of construction the star maps used, there is a little story behind it. Joseph Needham in his book claimed multiple times that the maps (and the ones in Su Song's *Xin Yixiang Fayao* [Essentials of the Method for the New Armillary Sphere and Celestial Globe]) utilized cylindrical orthomophic "Mercator" projection. Like his many other assertions, it was repeatedly quoted and followed by scholars at home and abroad — although Needham had not provided any evidence. However, in recent years, new research outcomes showed through calculations that these star maps were in no way drawn through application of the cylindrical orthomophic "Mercator" projection.[5]

P. 2512 and S. 3326 were highly valued by our knowledgeable predecessors. This is good, but what is notable is that no one seems to raise particularly the doubt that even though manuscripts of this sort are treasured today, they were not necessarily considered of high quality during the times they were written — since the preserved Dunhuang manuscripts have much to do with happenstance. Hence, the two pieces cannot be hailed as representing the highest attainment of *tianxue* then.

---

[4] Xi Zezong: "Dunhuang Star Map", *Historical Relics*, Issue 3 (1966).
[5] Hu Weijia: "Study on 'Escapements' and Star Maps Construction in *Xin Yixiang Fayao*", published in *Studies in The History of Natural Sciences*, Vol. 13, No. 3 (1994).

3. *Ershiba Xiu Ci Wei Jing* (*Twenty-eight Mansions Catalogue*)

The *Twenty-eight Mansions Catalogue* in P. 2512 is a complete work. It seems to also strictly follow the traditional data: the equatorial spans of each mansion (the degrees each mansion span along the equator) are identical to that recorded in *Tian Wen Xun of Huai Nan Zi* (*Patterns of Heaven, Huainanzi*) and *Lü Li Zhi of Han Shu* (*Rhythm and the Calendar, Book of Han*). On this, several studies have already been conducted. Pan Nai summarized in *The History of Fixed-Star Observation in China* as follows:

The "*Twenty-eight Mansions Catalogue*", as astronomical materials ... consists of observations made in 450 B.C. and 200 A.D. ... The *Twenty-eight Mansions Catalogue Twenty-eight Mansions Catalogue* itself constitutes part of *Shishi Xingjing*.[6]

Yet new research findings beg to differ. Because the question is closely related to *Shishi Xingjing*, and given that Hu Weijia has already demonstrated the much later birth of the star manual, he would certainly impugn the statement that *Twenty-eight Mansions Catalogue* can be dated to the Warring States Period or the Qin and Han *dynasties*. According to Hu:

The location of the twenty-eight Lunar Mansions recorded in *Kai Yuan Zhan Jing* is observed in 713 A.D. or shortly after. And the earliest *qujidu* table of the twenty-eight Lunar Mansions — the *Catalogue* — preserved till this day should be observed in the early-Tang or some time earlier.[7]

This assertion is naturally in line with his opinion of *Shishi Xingjing*. Besides, there are two notable works appearing between the *Twenty-eight Mansions Catalogue* and *Kai Yuan Zhan Jing*, namely, the *Tianwen Yaolu* written by Li Feng in 664 and the *Tiandi Ruixiang Zhi* by Sa Shouzhen in 666. The two works both have mere fragmentary copies preserved in Japan. Academician Xi Zezong has studied the copies and arrived at the conclusion that by virtue of the two works, *Twenty-eight Mansions Catalogue* successfully evolved into (the star manuals of the

---

[6] Pan Nai: *The History of Fixed-Star Observation in China*, (1989), p. 97–98.

[7] Hu Weijia: "Study on 'Escapements' and Star Maps Construction in *Xin Yixiang Fayao*", published in *Studies in The History of Natural Sciences*, Vol. 13, No. 3 (1994).

three schools in) *Kai Yuan Zhan Jing*.[8] Hu Weijia agrees with him on this point.

4. Chen Zhuo — Three colors for the three star maps

Pan Nai has devoted generous space in his monograph to the description in P. 2512 of the star manuals composed by the three authorities — Shi, Gan, and Wuxian. He mainly carefully tallied and compared the number of constellations and stars among the three schools; as a result, he finally counted 283 asterisms and 1464 stars (or 1465, due to varied treatment of the star "Shengong").[9]

Actually, problems around the three schools are also to be reckoned with in the history of Chinese astronomy, so they deserve some deliberation here.

Ancient Chinese astrology witnessed the rise of different schools, each with their own line of development. Judging from materials available now, at that time Shi, Gan, and Wuxian all had their own star manuals and star maps, and the stars their divinations bore on were also different. But the original materials in this regard are lacking.

Kai Yuan Zhan Jing preserved the information of the three astrological schools. As was suggested before, the star catalogues are likely from later observations, but Gan and Shi, as astrologists, were absolutely real historical figures, though placing Wuxian as such might be a bit tricky.[10] Pan Nai even thought that the so-called Wuxian stars were but the work of Chen Zhuo himself under the cloak of Wuxian. Such a saying is interestingly well-founded.[11]

---

[8] Xi Zezong: The Astronomy in Dunhuang Manuscripts, Report of the Fourth Chinese History of Science International Seminar, Sydney, (1986).

[9] Pan Nai: *The History of Fixed-Star Observation in China*, (1989), pp. 99–100.

[10] For research on the ages in which Shi, Gan and Wuxian lived, refer to pp. 77–89 of *Tianxue Zhenyuan*. Shi Shen (or Shi Shenfu) was born in the Warring States Period, and Gan De between the late Warring States Period and the Qin and Han *dynasties*. Wuxian was originally a famous *shaman* in the Shang Dynasty under the reign of King Taiwu; he then became the embodiment of ancient wuxi astrology, just like Qi Bo became the byword for the art of healing and Bian Que, Hua Tuo for famous healers in later generations. Hence, Wuxian also led a school of astrology.

[11] Pan Nai: *The History of Fixed-Star Observation in China*, (1989), pp. 115–117.

Chen Zhuo is a critical figure in the history of the three aforementioned star manuals and star maps. The dates of his birth and death cannot be traced, and no biography of him is included in the official history. But according to *Astronomical Treatise, the Book of Jin* and *Astronomical Treatise, the Book of Sui*, he was originally a court historian of the Eastern Wu. After Wu's demise was brought by Western Jin, Chen, like many in the upper class of Eastern Wu, took office in the court of Western Jin, serving once again as court historian during the reign of the Emperor Wu of Jin. This detail is enough to prove his status in the astronomical community. The Western Jin court had received talents in this field from both Cao Wei and Shu Han, so by no means could Chen rise to his position in the new dynasty without impressive accomplishments to his credit. Naturally, the later Yongjia Southward Migration must have felt like a revisit to the old place for him — the Western Jin collapsed in 317 A.D., followed by the Eastern Jin Dynasty established by Emperor Yuan of Jin who took the throne in Jiankang (present-day Nanjing) and whose enthronement day selection was participated by in Chen Zhuo in his capacity as the court historian. Afterwards, the historical literature lost track of him. During his tenure in the Eastern Wu, a remarkable feat Chen Zhuo achieved was consolidating the work of the three astrological schools.

Tian Wen Zhi of Sui Shu (Astronomical Treatise, the Book of Sui) says:

During the Three Kingdoms period, court historian Chen Zhuo embarked on categorizing the asterisms of the three schools — Gan, Shi and Wuxian; he recorded them with pictorial explanations and divination interpretations, and added zan (a style of writing). In total, he counted 254 asterisms, including 1283 stars which, factoring in the 28 mansions and subsidiary asterisms with 182 stars, accumulate to 1465.

Ever since Chen Zhuo combined the stars of the three schools, problems ensued about how to distinguish them in a general star map. The solution that occurred in the mind of the ancients was to use different colors to flag the stars of the three different schools. This should have fared well, but due to contradictions arising in later generations, a series of "color events" were triggered. The first adopter of the solution was Qian Lezhi, court historian of the Liu Song Dynasty, who fashioned a

bronze armillary sphere. *Astronomical Treatise, the Book of Sui* described
this as:

> *During the Yuanjia Era of the Song Dynasty, court historian Qian Lezhi
> made a bronze armillary sphere which used red, black and white to
> mark the stars of the three schools, their total number identical to Chen
> Zhuo's count. ... (Emperor Wen of Sui) ordered Yu Jicai and some oth-
> ers to correct the size, distribution and location of the stars of the three
> schools, with reference to the old private and official maps from dynas-
> ties including the Zhou, Qi, Liang, Chen, and personages including Zu
> Geng and Sun Senghua, in an effort to make a star map based on the
> theory of canopy-heavens.*

Not surprisingly, Qian's bronze armillary sphere failed to survive the
test of time. Though quite a few star maps were preserved in *Lingtai
Miyuan* written by Yu Jicai, its available version now is the result of a
Northern Song revision by people like Wang Anli; therefore, it is hard to
say whether those maps belong to the Northern Song or the period
between the Sui and Zhou *dynasties*, or an age much earlier.

Qian Lezhi distinguished the three schools with red, black, and white;
this surely worked on bronze ware, but when it comes to walls, silk, or
paper, the white color will blend into the background, which necessitated
workarounds. In the tomb of Feng Sufu, a Northern Yan official, a stone
coffin was found with the top covered by a star map (just a picture of
stars — not presenting the accurate location of stars) in red, yellow, and
green; in the tomb built for crown prince Zhanghuai of the Tang Dynasty,
the back chamber featured a star-map top finished with golden foil, silver
foil, and yellow paint; besides, in Manuscript P. 2512, there are specific
records:

> *Shi's star map has in its center 64 constellations including 270 stars,
> shown in red, and on the periphery 30 constellations including 257
> stars...shown in red; Gan's star map has 42 constellations including
> 230 stars on the periphery... shown in black, 76 constellations including
> 281 stars in the center, all shown in black; Wuxian's star map has 44
> constellations including 134 stars in total, shown in yellow.*

By the same token, *Xuan Xiang Shi* (incomplete) in Manuscript P. 3589 marked the stars mapped by the three schools (words "red", "black", and "yellow" were prefixed, respectively, to the names of the stars attributed to Shi, Gan, and Wuxian). Though P. 2512 and S. 3326 contain no star maps, the latter gives a clue as to the different symbols used for stars — black dots for Gan's stars and red-filled black circles for Shi's and Wuxian's stars. Moreover, a star map handed down through the Dunhuang manuscripts, often called "Ziweiyuan Star Map" (Item No. 58 of Dunhuang Museum collection), also adopted the same symbols.

Besides, the historical literature also offers varied records of the representative symbols, like the star map in *Xin Yixiang Fayao* written by Northern Song author Su Song, the Japanese star chart Koshi Tsuki Shinzu (dated to about 1100 A.D., regarded as the oldest star map of Japan), the story of Xu Ziyi's examination in the "Cixue" section, Volume One, of *Sichao Jianwen Lu* written by Southern Song author Ye Shaoweng, etc. The different color choices for the three schools' stars and their corresponding historical sources are tabulated below.

	Shi	Gan	Wuxian
The bronze armillary sphere of Qian Lezhi in *Astronomical Treatise, the Book of Sui*	Black	Red	White
Dunhuang Manuscript P. 2512	Red	Black	Yellow
Dunhuang Manuscript P. 3589	Red	Black	Yellow
Dunhuang Manuscript P. 3326	Red	Black	Red
Dunhuang "Ziweiyuan Star Map"	Red	Black	Red
Volume 2 of Song Susong's *Xin Yixiang Fayao*,	Red	Black	Yellow
Koshi Tsuki Shinzu	Red	Black	Yellow
"Cixue" section, Volume One, *Sichao Jianwen Lu* by Southern Song author Ye Shaoweng	Black	Red	Yellow

The color-based differentiation of stars is a remnant from early astrological schools. After Chen Zhuo integrated the three schools, it made no sense from a pragmatic point of view, so it is only natural that the later star maps gradually abandoned this unnecessary labor.

5. *Xuan Xiang Shi* (*Poem of the Occult Images*), *Bu Tian Ge* (*Song of the Sky Pacers*), and other related works

The *Xuan Xiang Shi* in Manuscript P. 2512 is a popular work describing the stars in the form of a poem; though called a poem ("shi" means poem in Chinese), it is actually of no literary value. Here are several lines extracted from its beginning and end:

> *The Horn mansion, the Neck mansion and the Root mansion form a straight line from east to west. The Arsenal asterism lies in the south of the Horn mansion; the Judging lies to the north of the Arsenal; and below the Arsenal is the Southern Gate. In the south of the Root mansion is the Imperial Guards, on whose north until the handle of the Big Dipper lies respectively the Conductors, the Great Horn, the Celestial Lance, and the Twinkling Indicator...The Big Dipper is not included in the poem because everyone knows its location. The Kui and Lou mansions lie due north, and the Zhen and Yi mansions shine due south. Following this rule, which constellation can hide from plain sight?*

Manuscript P. 3589 also contained the *Xuan Xiang Shi*, but it is incomplete and conflicting with P. 2512's version; however, it has an inscription stating "Written by Court Historian Chen Zhuo", according to which Pan Nai attributed the poem to Chen Zhuo.[12] Still, a close look at those simple and exoteric lines leads one to doubt if they seem at odds with such a personage as Chen Zhou in that particular period of time — it's only a doubt of course.

Compared with *Xuan Xiang Shi*, another work of the same type is much more important, that is, *Bu Tian Ge* (Song of the Sky Pacers).

Bu Tian Ge has been passed down to today's audience. The "Astronomy" section in *Treatise on Literature, New Book of Tang* (*3*) recorded that "Wang Ximing, Danyuanzi, *Bu Tian Ge* (*Vol. 1*)"; in addition, books including Zheng Qiao's *Tongzhi*, Chen Zhensun's *Zhizhai Shulu Jieti*, Chao Gongwu's *Junzhai Dushuzhi,* and the like also listed the poem. But throughout history, dispute never ceased over whether Wang Ximing lived in the Sui Dynasty or the Tang Dynasty, and whether he and Danyuanzi are the same person.

---

[12] Pan Nai: *The History of Fixed-Star Observation in China*, (1989), p. 133.

In my humble opinion, Professor Chen Shangjun's investigation is the most reliable, and the text he collated is also preferable to other available versions.[13] In his words, Wang Ximing is a figure of the Tang Dynasty, had served as Youshiyi (Right Adviser) during the Kaiyuan Era, and had written the ten-volume collection *Tai Yi Jin Jing Shi Jing*; "Danyuanzi" should be his *hao (style name)*, or penname.

Bu Tian Ge described the 283 asterisms including 1,464 stars tallied by Chen Zhuo in heptasyllabic verse. It is obviously more literarily refined than *Xuan Xiang Shi*. Take "Ox", one of the seven northern mansions for example:

> *The six stars of the Ox lie near the Milk Way. Though with two complete horns, the Ox is forever a leg short. Below it spread the nine stars of the Celestial Farmland and below the Celestial Farmland, three trios of stars link to form the Nine Water Wells. Over the Ox stand three River Drums over which are three stars called the Weaving Girl. The Left Flag and Right Flag both consist of nine stars; they flank the River Drums but the former shines brighter. What's more, there's a four-star asterism named the Celestial Drumstick, lying right below the River Drums like a string of pearls; the Network of Dykes is located east of the Ox. The Clepsydra Terrace resembles an open mouth, and the Imperial Passageway connects five stars. Where are the two asterisms? To find them you look at the vicinity of the Weaving Girl.*

The greatest significance of *Bu Tian Ge* is that it is by far the earliest historical document specifying the division of the starry sky into "Three Enclosures and Twenty-eight Mansions". The Twenty-eight Lunar Mansions have long existed — their origin is a very troubling but intriguing topic that we will mention again later. The Three Enclosures refer to the Purple Forbidden Enclosure, the Supreme Palace Enclosure, and the Heavenly Market Enclosure. They began to take shape in *Tianguan Shu of Shih Chi*, their name was already known in *Xuan Xiang Shi*, but they were not truly established until *Bu Tian Ge*, from which time they remained in

---

[13] See *Quan Tangshi Bubian (Complemented Version of Complete Tang Poems)* edited by Chen Shangjun, Zhonghua Book Company, Vol. 2, (1992) edition, pp. 805–811.

use for as long as 1,200 years or so. In the late Ming period, Jesuits spread the Western astronomy to China; the Qing Dynasty, though making European astronomy the theoretic base for official astronomy, merely catalogued the English equivalents of the Chinese stars. It was not until the 20th century that China fully embraced Western astronomy, abandoning the traditional divisions, namely, the "Three Enclosures and Twenty-eight Mansions".

Besides *Xuan Xiang Shi* and *Bu Tian Ge*, other works on astrological phenomena were also seen before and after their times, such as *Tianxiang Fu* (*Rhapsody on Astronomical Phenomena*), which was said to be written by Zhang Heng of the Eastern Han Dynasty but was unfortunately lost; *Guanxiang Fu* (*Observations Rhapsody*) written by Zhang Yuan in the Northern Wei Dynasty under Emperor Taiwu's reign; *Tianwen Daxiang Fu* by Li Bo (father of Li Chunfeng) of the Sui and Tang Dynasties; *Huntian Fu* by Yang Jiong, one of the Four Paragons of the Early Tang; *Xing Fu* (*Star Rhapsody*) by Wu Shu of the Song Dynasty; *Zi Wei Yuan Fu* by Wang Kekuan of the Yuan Dynasty; and *Xingxiang Fu* (Rhapsody on Astrological Phenomena) by Wu Xi of the Qing Dynasty. Among these works, except *Tianwen Daxiang Fu*, which is fairly professional, others are just a literati's play with words — at best second- or third-tier literary works — and should not be deemed rival to the quite specialized *Xuan Xiang Shi* and *Bu Tian Ge*. Zhen Qiao's exclamatory statement about *Bu Tian Ge* in *Tongzhi, Treatise on Astronomy* offered some clues in this regard:

> *I had ever read through all the relevant books, but to no avail; and I had also searched for any helpful star maps, still without accurate information. One day I got the Bu Tian Ge and began to read it. It was a moonless autumn night with a clear sky. I stared up at a star in the pauses between each two sentences; within a few days I successfully remembered all the stars in the sky! The book is circulated among observatories only, not accessible to the common multitude. Astrologists keep it secret and call it "Guiliaoqiao".*

In other words, *Xuan Xiang Shi* and *Bu Tian Ge* are "the secret texts of *tianxue*", but those rhapsodies are not.

6. Mawangdui silk manuscript *Wu Xing Zhan* (*Divination of the Five Planets*) and *Tianwen Qixiang Zazhan* (*Divination by Astrological and Meteorological Phenomena*)

Important secret texts of tianxue were also discovered amid the manuscript horde at Mawangdui.

Mawangdui silk manuscript *Wu Xing Zhan* (*Divination of the Five Planets*). It is known as the oldest dated text so far in Chinese astrological literature. The text provides 70-year visibility tables for Venus, Saturn, and Jupiter, running from 246 B.C. to 177 B.C. Because its contents include both general and specific discussions on the five planets (Jupiter, Mars, Saturn, Mercury, and Venus) and yet the visibility tables cover only three, the manuscript is probably incomplete.

The general and specific discussions on the five planets are all about astrological theories. For example, in *"General Discussion on Five Planets"* (the subtitles are all added by modern compilers), it is said:

When Venus appears over a country, its color should be observed daily. During the day, observers can see its beautiful color. If it shows a beautiful color, then the country is promised victories in war. If it's lost from sight, then the country has weak military strength; if such a phenomenon occurs three times, then the country can be defeated and its leaders can be seized.

Another example comes from the "Mars" chapter:

When Mang (ray) meets the Heart Mansion, there will be funerals; no matter it appears to the south or north of Xinxing, deaths are presaged. If it is red, then the southern countries will have luck; if it's white, the countries in West will have luck; if it's black, the northern countries will have luck; if it's green, the eastern countries will have luck; and if it's yellow, the central countries will have luck.

Chapters on other planets are all similar. But through these astrological descriptions, we can see from *Wu Xing Zhan* the astronomical accomplishment at that time. According to the research by Academician Xi Zezong, the position tables of the three planets — Saturn, Jupiter, and

Venus — were derived from calculation based on the observational data in 246 B.C. and the planetary cycles known in the Qin and Han Dynasties.[14] The manuscript specified the synodic periods[15] of the three planets.

Venus: 584.40 days; Jupiter: 395.44 days; Saturn: 377 days. Here, let us compare the synodic periods adopted by *Wu Xing Zhan*, the ancient Babylonians, and modern astronomy. See the chart below (unit annualized) for further details.

	Wu Xing Zhan	The Ancient Babylonians	Modern Astronomy
Venus	1.600	1.599	1.599
Jupiter	1.083	1.092	1.092
Saturn	1.032	1.035	1.035

The values adopted by *Wu Xing Zhan* and modern astronomy are quite close.

Mawangdui silk manuscript *Tianwen Qixiang Zazhan*. This is another early classic astrological work in the Mawangdui cache of silk manuscripts. The text contains over 350 entries of divination, each featuring a picture in red, or black, or red and black followed by a title, an explanation, and divination statements. Take several divination statements as examples, as given in the following:

> *With red clouds as such, there will be war in the second lunar month. With a comet as such, there will be new weapons to fill the arsenal and a grand harvest. If the moon blots out a star, then a country is facing its demise. If the star comes out, then the country will reclaim its independence; but if not, the country fails out and out.*

The 29 comet drawings in the manuscript have gained great fame at home and abroad as scholars carried out studies on a large scale; they are

---

[14] Xi Zezong: "An Important Finding in the History of Chinese Astronomy — *Wu Xing Zhan* from a Silk Manuscript Horde in A Mawangdui Han-dynasty Tomb", the first essay of A Collection of Essays on History of Chinese Astronomy, Science Press, (1978).

[15] Synodic period, the time required for a planet to return to the same elongation moving in the same direction. Here the "elongation" refers to (the projection on the ecliptic of) the geocentric angle between a planet and the sun.

often referred to as an independent collection called *"Huixing Tu (Charts of Comets)"*. Actually, the manuscript is just a work on divination from cloud shapes and astronomical phenomena, without any scientific intention in the modern sense.

The Section "Divination by Cloud Shapes" in *Tianwen Qixiang Zazhan* belongs to the early literature of a long-held tradition and deserves special attention. Take the first five items in the first column for example:

Text Pattern
Clouds in Chu are white, like the sun Sun
Clouds in Zhao Bull
Clouds in Zhongshan Bull
Clouds in Yan   Tree
Clouds in Qin   Woman

To see bulls, trees, and women in clouds surely requires imagination and association, but it is not the short-lived fantasy of a writer with artistic tendency to match the clouds of states with all kinds of patterns; rather, it is a long-held tradition in Chinese astrology. For instance, we can see in *Astronomical Treatise, the Book of Jin* such accounts:

> *The clouds are like cloth in Han, bulls in Zhao, the sun in Chu, carriages in Song, horses in Lu, dogs in Wei, wheels in Zhou, pedestrians in Qin, mice in Wei, crimson robes in Zheng, dragons in Yue and barns in Shu.*

Though it has not been mentioned "the clouds in Zhongshan", "the clouds in the state of Yan", and the statement that "the clouds in Qin are like pedestrians" are inconsistent with the former analogy to women, records show that the eighth volume of *Taiping Yulan (Readings of the Taiping Era)* has quotes from *Bingshu* saying that "the clouds of Qin are like beauties".

Then let us look at the fifteenth item of the first column, which presents a fish pattern captioned "heavy rain". What is the relation between fish and heavy rain? After consulting the forty-ninth volume of *Kai Yuan*

Zhan Jing which recorded that "Pre-rain clouds are like fish and dragons, with a greenish and watery color", we can figure out that the divination in *Tianwen Qixiang Zazhan* means fish-shaped clouds foretold of heavy showers. The above examples show that *Tianwen Qixiang Zazhan* can indeed be seen as the origin of many traditional astrological sayings in the later generations.

7. *Yu Li Tong Zheng Jing* and *Qiankun Bianyi Lu*

In the last two sections of this chapter, we will again return to the astrological pieces with available texts — though they are not as important as those discussed in the previous chapter. Besides the already introduced *Yisi Zhan*, several other astrological works attributed to Li Chunfeng have been handed down to us.

Yu Li Tong Zheng Jing, three volumes, all handwritten copies, inscribed "Compiled by Tang-dynasty Grand Master Li Chunfeng". The first volume deliberates on the sky, clouds, rain, qi, rainbow, wind, thunder, mist, frost, snow, hail, graupel, dew, morning/evening glow, earthquake, flood, draught, conflagration, etc. The second volume discusses the sun, the moon, Jupiter, Mars, Saturn, Mercury, Venus, meteors, ominous stars, the ecliptic, and *fenye*. The last volume focuses on the Three Enclosures (the Purple Forbidden Enclosure, the Supreme Palace Enclosure, and the Heavenly Market Enclosure), the asterisms of the Twenty-eight Lunar Mansions, and divination. The discussions do not go beyond the typical astrological boilerplate and are laden with moral indoctrination, which can be evidenced by the following extract from the first volume:

> *The vagaries of the heaven are like fatherly rage. Fathers are much-respected and strict; they often scold their sons. If they do not, their sons will run amok. The heavenly changes forebode disasters, of which the rulers have to be afraid. The vagaries of the earth are like motherly rage. Mothers are kind and loving; they seldom lose temper. Earthly changes warn the rulers of their fate. Precautions can be deployed against heavenly changes which are mostly mere admonishments, but earthly changes leave no way out, because she's seldom given to anger.*

At the end of the book is an epilogue with the same inscription "Compiled by Tang-dynasty Grand Master Li Chunfeng". It notes the standard of behavior for astrologists and is quite interesting:

> *God is just and infallible. But misinterpretation or incomprehensive observation can lead a prediction to fall through. If one can divine the mystery to know the changes and see through the astronomical phenomena to gain insight, then every prediction will be born out. The practice of divination is never about pursuit of good auspices; rather, it's for the avoidance of disasters, so even in a time of peace and prosperity, it's necessary to alert the throne of the bad omens divined from astronomical phenomena, so that the rulers can be prepared for danger. If misfortunes befall a country, it's important to read the astronomical phenomena based on divination principles to seek hope and avoid the doom. This is what we say to look out for danger in peace and seek propitious signs in crisis.*

The general idea is in line with the opinion frequently expressed by Li Chunfeng in other places. On the whole, the book is like an abridged version of *Yisi Zhan*, so it is not unreasonable to attribute the book to Li Chunfeng.

Qiankun Bianyi Lu (*Records of Changes of the Universe*), no volumes contained, a handwritten copy, inscribed "Compiled by Tang-dynasty Grand Master Li Chunfeng". Its contents, simple and shallow, also revolve around common astrological topics; it is hard to imagine that the author of *Yisi Zhan* could have written such a book. It could be the work of an imposter from the later generations.

In addition, the book *Gai Zheng Guang Xiang Wan Zhan* (*Divinations on Celestial Observations*) is sometimes subsumed under Li's oeuvre. It is a ten-volume collection of handwritten copies and its contents are as follows:

Volume 1: Divination on the Sky, Divination on the Sun, Divination on the Moon
Volume 2: Divination on the Five Planets
Volume 3: Purple Forbidden Enclosure, Supreme Palace Enclosure, Heavenly Market Enclosure

Volume 4: Seven Mansions in the Azure Dragon of the East
Volume 5: Seven Mansions in the Black Turtle of the North
Volume 6: Seven Mansions in the White Tiger of the West
Volume 7: Seven Mansions in the Vermilion Bird of the South
Volume 8: Divination on Other Stars, Divination on Weather
Volume 9: Fengjiao ("Wind Angles" divination)
Volume 10: Divination on Astronomical Phenomena

The book's structure is very different from *Yisi Zhan*. And the year of its publication is difficult to identify — as the ancient Chinese literature in astrology (and many other fields as well) evolved along a lineage, that is, most achievements had drawn widely on the works of the predecessors, it is difficult to trace from the divinations the gradual changes over time. In my humble opinion, *Gai Zheng Guang Xiang Wan Zhan* should be dated much later than *Kai Yuan Zhan Jing* and *Yisi Zhan*, so I'm afraid that attributing it to Li Chunfeng is hardly convincing.

8. Two *Star Manuals*

A few short pieces that have come down to us but remain absent from common monographs on the history of astronomy are also be briefly introduced in what follows.

Xing Jing (Star Manual). It has an inscription that says: "written by Master Gan and Shi Shen of Han Dynasty, collated by Li Rong of Nanchang". The book contains two volumes and describes in total 167 asterisms. Each asterism is first given a simple illustration, then a brief introduction of its location, its astrological meaning, and finally relevant divination methods. Here are two examples:

> *The Tianyi Star lies south of the left star outside the gate of the Purple Forbidden Enclosure. It symbolizes God, and predicts about wars. If the star is bright, it bodes well; if the star is dim, it bodes ill. If the star leaves its original place and approaches the Dipper Mansion, within ninety days there will be war. When the star shines brilliantly, yin and yang are in harmony, everything thrives, and the king is blessed; should the star disappear, the world will plunge into chaos, and great disasters will befall.*

The Taiyi Star is half degree south of the Tianyi Star. It also symbolizes God and heads sixteen deities. The star forebodes wind, rain, floods, droughts, war, famine, diseases and disasters for a country. If the star is bright, it bodes well; if it's dim, the opposite is true. If it leaves its original place and approaches the Dipper Mansion, within ninety days there will be war.

The book has been included in *Congshu Jicheng (Complete Collection of Books from [Various] Collectanea)*, but the publication is full of errors. The inscription "written by Master Gan and Shi Shen of Han Dynasty, edited by Li Rong of Nanchang" is obviously doubtful. And its postscript written by Wang Mo indicates that Wang also impugned the book's authorship, though he acknowledged its possible connection to Gan and Shi.

Tong Zhan Da Xiang Li Xing Jing. It was originally subsumed under the Zhongshu Category, Dongzhenbu Volume of *Dao Zang* (Depository of Taoist Works), and has also been included in *Congshu Jicheng*; the book has no inscription announcing authorship. Its contents are similar to that of the aforementioned *Xing Jing*, also introducing 167 asterisms but in a different order. A brief glance is enough to see that they either have identical authorship or are along the same lineage.

8.1. *Yunqi Zhanhou Pian and Tianwen Zhanyan*

Yunqi Zhanhou Pian. The book has an inscription that says: "written by Tao Luzi"; it contains two sections and has been included in *Congshu Jicheng*. The whole book, written in the style of verse, revolves around the divination of *chi*:

The qi of true lords is like a barn or a gate tower, perfect with a yellow or red color. If purple qi appears above an army, the army is hard to resist. If purple qi takes the shape of a rainbow, it means the Big Dipper is presenting a good omen. If purple qi takes the shape of an umbrella, great things are happening below. If green qi turns purple, it predicts unfathomable promise.

And of course, it also has, in rewritten form, in the verse style the part in *Astronomical Treatise, the Book of Jin* that describes the clouds of states.

Tianwen Zhanyan. It is of unknown authorship, with an inscription that says: "collated and sent to press by Zhou Lüjing", and it is also in the *Congshu Jicheng* collection. The book mainly collects the popular weather lore sayings with vernacular wording and precariously founded predictions. For example, "if the west wind blows in the first day of the first lunar month, it means the burglars will run rampant, and if it's accompanied by heavy snow, then with it evil spirits will follow; if the winter solstice remains overcast all day, the next year is blessed with peace".

Astronomy and metrology are two different disciplines according to modern classification, but the "astronomy (*tianwen*)" the ancients referred to actually included today's metrology, so I explained it here for the record.

Chapter 7

Universe in the Eye of Ancient Chinese

1. Tune: The Magpie on the Bough

Her passionate heart I've already won.
Mighty the river tides.
Which fail to have my love-bridge done.
My dreams frequent her starry dwelling place.
But time-space keeps in this world the waked man.

Curtains of twelve doors veil the fair-faced one.
Deep the celestial main.
Deeper my tender heart has run.
To long t' reencounter her in flower shade.
I'll tread all recesses trodden by none.

This poem with the tune *Que Ta Zhi* (*The Magpie on the Bough*) was written by the physics professor Ge Ge on the same lines as one of the famous poems of Feng Yansi (903–960) in the Five Dynasties Period. It is good enough to pass for a genuine classic poem. The line "But time-space keeps in this world the waked man" shows his understanding of electrodynamics and the theory of relativity. The concept time-space comes from the West. Ancients in China never used this word.

1.1. *The definition of the universe*

According to *Shi Zi*, which is believed to be written in Han Dynasty,

> *Yu is made up of north, south, east and west and Zhou is about the time beginning from ancient times.* (*The two make Yuzhou-the Universe.*)

This is the concept that best suits modern time-space in ancient Chinese books. But for this, we should never assume that the author (Shi Jiao, in Zhou Dynasty) was a materialist, because he gave many idealist remarks like *"The Sun has five colors, representing the ultimate yang essence, symbolizing the virtues of the Emperor"*. With the five-color sun shining, the Emperor took to the throne.

Nowadays, Yuzhou (universe) sounds very easy to understand as the word merely means space, heaven, and earth, which is what the ancients used very often. On the other hand, time-space sounds a little academic, but it is an easy-to-understand expression these days. This is truly an interesting usage of language.

When talking about theories of universe in ancient China, many previous works focused on six schools of theories about the universe, including Gaitian, Huntian, Xuanye, Xintian, Qiongtian, and Antian. According to Jin Shu-Tian Wen Zhi (*Astronomical Treatise, the Book of Jin*), there are only three major schools of astronomy, namely, Gaitian, Xuanye, and Huntian.

The discussion now will begin with a description of physical properties of models of the universe from the perspectives of the abovementioned three schools. At the end of the day, we will discuss outlooks of the universe considering the presence and actions of people. In this case, ancients refrained from using Yuzhou as their choice of word, but they somehow echoed with modern cosmology.

2. Is the Universe finite or infinite?

Now that Yu means space and Zhou stands for time, is there a boundary for time-space? Is there a beginning and an end of time? They are naturally common-sense and logical questions. But they have baffled people both today and in the past.

Nowadays, people are baffled because many are firm believers of Engels, a saint, who alleged that the universe is infinite. As a result, they believe in the infinity of the universe. But his conclusion was drawn well before scientific evidence about cosmology was discovered. What the saint said was a mere outcome of his own thinking instead of being based on well-grounded evidence like red shift, 3K background radiation, and helium abundance. Further, what is to be made of thinking alone when scientific evidence has already made it clear.

Being puzzled, people have gone a step further by treating the ancients unfairly. Ancients that stood for a finite universe are accused of being idealistic and rebellious, whereas champions of infinity are crowned as being materialistic and progressive. In this sense, the presence of different thoughts among the ancients has been incorrectly labeled as a fight among them. This was especially true a little while after the Cultural Revolution (1966–1976). Till date, such a cliché is prevalent in the academic community.

Astronomers were among the first acceptors of modern cosmological evidence. A case in point is the establishment of the Big Bang model based on scientific observation. The model believes in the presence of a starting point of time and boundaries in space. If one has to make a choice between finitude and infinity, the former is the only option.

Forefathers had to think hard about the universe given that they had no evidence of modern cosmology. *Zhou Bi Suan Jing* (*The Arithmetical Classic of the Gnomon and the Circular Paths of Heaven*) clearly states that the universe is composed of two 2D circles 810,000 li (405,000 km) in diameter. Later, this book will prove the universe does not take the shape of two spherical crowns.[1] In his book *Ling Xian*, Zhang Heng said the heaven and earth is a globe 200,032,000 li (Note: one li is half a kilometer) in diameter. He added that

> *Beyond the scope is something I couldn't understand. It is the universe that I fail to understand because it is infinite.*

Zhang called what lies outside the heaven and earth as Yuzhou (universe). Unlike *Zhou Bi Suan Jing*, he clearly advocated the infinity of the

---

[1] This question will be dealt with in the next chapter. Chapter Nine will discuss a highly possible Indian origin of the universe model.

universe, which of course was also a mere result of his own thinking because he was unlikely to discover scientific evidence then. Even though his thought agree with conclusions of modern evidence, this is just sheer coincidence. After all, there were many thoughts conflicting scientific evidence.

Some argued for a finite universe, including Yang Xiong of Han Dynasty, who defined, in *Taixuan Xuanchi*, the universe as

> *Yu includes the heaven and what's under it while Zhou came into being when the heaven and earth was born.*

Yu includes the heaven and what is under it while Zhou came into being when the heaven and earth were created. In this sense, Yuzhou is clearly confined to be within the heaven and earth. Thinking with common sense, such a view can be readily acceptable to people who are not science-minded. So, more people believed in a finite universe as it is close to the heart of people, though some snippets of information from ancient works revealed the proposition for an infinite universe. *Tian Dui*, written by Liu Zongyuan in Tang Dynasty, is a case in point.[2]

Whether the universe is infinite or not was not an important question in astronomy, astrology, or philosophy for the ancient Chinese. The definition that Yu stands for space and Zhou for time is acceptable to finitude supporters, infinity champions, and those who believe it is impossible to know whether the universe is finite or not.

3. Exaggerated Xuanye theory

In *Mathematics and the Sciences of the Heavens and Earth*, the 3rd volume of *Science and Civilization in China*, Joseph Needham devoted a section to Xuanye theory. He passionately praised the universe model by saying:

Such a progressive outlook of the universe is as brilliant as any theory proposed in Greece. Crystal spheres put forward by Aristotle and Ptolemy

---

[2] Please refer to Zheng Wenguang & Xi Zezong. *Zhong Guo Li Shi Shang De Yu Zhou Li Lun* (*Universe Theories in Chinese History*). People's Publishing House, (1975). pp. 145–146.

dominated astronomy in Europe for more than 1,000 years. The Chinese Xuanye theory, which believes in sparsely floating celestial bodies in an infinite space, is much more advanced than the European theory. Though sinologists tend to believe that Xuanye theory hasn't even worked, it has made greater difference to astronomy in China than it seems.[3]

This brought much fame to Xuanye theory, which has since then been eulogized for being materialistic. People also praised the theory by saying that it came to light about many years before Giordano Bruno proposed anything. I pointed out over 10 years ago that two technical mistakes were made in this paragraph,[4] but they are minor issues. At issue is whether Joseph's comment on Xuanye theory is appropriate.

All the historic data about Xuanye theory can only be found in one paragraph quoted by Joseph Needham in *Tian Wen Zhi of Jin Shu* (*Astronomical Treatise, the Book of Jin*).

Books about Xuanye are already lost. Xi Meng, a librarian in Han Dynasty, was the only person that recorded what his teacher said: *The universe is without substances. It you look up to the universe, you cannot see its edge no matter how hard you try. Likewise, you probably believe that the Yellow Mountain is green if you view it from a long distance. If you take a bird's-eye view at a barranco, you see only black. So you never know what the true color it is. The sun, the moon and stars are all floating in the universe, driven by Qi, or energy. So there are no rules governing how the Sun, the Moon, Venus, Neptune, Mercury, Mars and Saturn advance or retreat. Unlike other stars that often disappear after they travel from east to west, the North Star's location remains unchanged. Sheti and Tianxing moves towards the east. They move one degree everyday and 13 degrees every month. Their movement is sometimes fast and sometimes slow because they have no basis.*

---

[3] Joseph Needham. Astronomy, *Science and Civilization in China*. Science Press, Vol. 4, (1975), pp. 115–116.

[4] Two technical mistakes of Joseph Needham include: (1) Ptolemy's universe model builds on geometrical forms only about how stars move instead of concrete entities like crystal balls; (2) as Aristotle's theory was not recognized by the church until the 14th century, crystal spheres could dominate Europe's ideology on astronomy only up to 400 years. Please refer to Crystal Spheres in the History of Astronomy. *Astronomy Journal*, Vol. 28. No. 4. (1987).

A close look at this paragraph would lead you to conclude that what Joseph Needham spoke highly of was built on his wishful thinking.

First, the paragraph says nothing about the infinity of the universe. It claims that the universe has no edge. But it actually means the naked eye has its limitations. Second, it also alleges that there are no rules governing how the Sun, the Moon, Venus, Neptune, Mercury, Mars, and Saturn advance or retreat.[5] But it fails to explain the reasons, simply attributing it to the proposition that the stars have no basis. This means that such a theory carries little positive prospective. In comparison, theories prior to those given by Copernicus and even the Crystal Sphere Theory of Aristotle could stand to scientific test. Xuanye Theory is the mere result of thinking although it is similar to how we perceive the universe today. But the Crystal Sphere and others are the outcome of science, though they are contrary to our understanding of the universe.[6] All in all, theories other than Xuanye prevail.

Although Xuanye Theory has enjoyed great reputation thanks to Needham's praise, it fails to consider even the most basic concept of mathematical astronomy, which is committed to interpreting and mathematically describing astronomical phenomena and calculating astronomical possibilities. In this sense, Xuanye Theory, let alone Xintian Theory, Qiongtian Theory, and Antian Theory, is remotely on a par with Theory of Canopy Heavens and Huntian Theory, which indeed made a huge difference in ancient China. The following sections will discuss Gaitian (theory of canopy heavens) and Huntian (theory of sphere heavens) theories.

4. Self-conflicting theory of canopy heavens in *Zhou Bi Suan Jing* (*The Arithmetical Classic of the Gnomon and the Circular Paths of Heaven*)

According to the universe model of Gaitian from *Zhou Bi Suan Jing* (*The Arithmetical Classic of the Gnomon and the Circular Paths of Heaven*),

---

[5]When the Copernican theory came to light, the Ptolemy's system was much more accurate than the Copernican theory. This is because Tycho brought out the full potential of the Ptolemy's system.

[6]The so called "empirical" means it is based on scientific observations. According to the theories of modern science and philosophy, such a theory is "scientific".

the modern academic community agreed on the shapes of both heaven and earth, which are best represented by the following:

The book also says that "*the heaven resembles a covering rain hat, the earth resembles a dish upside down*". The celestial north pole is 60,000 li higher than the heaven of the Tropic of Capricorn, which is 20,000 li higher than the earth's surface of the celestial north pole. Likewise, the earth's surface of the celestial north pole is 60,000 li higher than that of the Tropic of Capricorn.[7]

However, it also says that the above-mentioned presumptive shapes of the heaven and earth are contradictory to the presumptive calculations in *Zhou Bi Suan Jing*. An example in this regard is given in what follows.

In Volume 2 of *Zhou Bi Suan Jing*, Chenzi said that the heaven is 80,000 li higher than the earth based on the presumption that the earth is flat. This is in conflict with the assumption that north pole's ground surface is 60,000 li higher than that of Tropic of Capricorn because the theory is that one does not believe that the earth is flat but argues that it is shaped like an upside-down plate.[8]

As early as the Tang Dynasty, Li Chunfeng already pointed out the self-conflicting nature of the theory on the shapes of heaven and earth from *Zhou Bi Suan Jing*. Li believed that such a theory was a blunder due to its self-contradictory nature.[9]

That said, all the above-mentioned books made a host of mistakes, including misunderstanding of what is said in *Zhou Bi Suan Jing*, credulity of the forefathers' conclusions, and failure to dig deeper. But why did they misinterpret what is said in the book? This is because of a lack of understanding of some key points in *Zhou Bi Suan Jing*, with the North Star being one of them. The next section will discuss Beiji Xuanji before analyzing what has been misunderstood.

---

[7] Bo Shuren, Another Discussion of the Theory of Canopy Heavens from *Zhou Bi Suan Jing* — in Memory of the 15th Anniversary of Mr Qian Baozong's Death. *Studies in the History of Natural Science*, Vol. 8. No. 4 (1989). The opinion echoed with what was said by Qian Baozong and Chen Zunwei.

[8] Chen Zungui. Zhong Guo Tian Wen Xue Shi (*History of Astronomy in China*). Shanghai People's Publishing House, Vol. 1 (1980), p. 136.

[9] *Zhou Bi Suan Jing*, one of the ten *Suan Jings* collated by Qian Baozong. Zhonghua Book Company, (1963), p. 28.

5. What is Beiji Xuanji?

The North Star holds one of the keys to figuring out what the heaven and earth look like in the Gaitian universe model. No consensus was reached as to what exactly Beiji Xuanji was among existing studies. Qian Baozong agreed with Gu Guanguang, who believed that "Beiji Xuanji" was a virtual star instead of a real one.[10] Chen Zungei stated expressly that Beiji Xuanji refers to the then North Star (Polaris). Beiji Xuanji in *Zhou Bi Suan Jing* means the big star in the North Pole, which is supposed to be between kochab and Star β of Ursa Minor, according to historic evidence and astronomical calculations.[11]

Beiji Xuanji or Xuanji is mentioned no fewer than three times in *Zhou Bi Suan Jing*, but the above quotation refers to only one of them. Many authors refrain from talking about the others, because it is impossible to interpret other discussions about Beiji Xuanji in *Zhou Bi Suan Jing* due to the preconception that both heaven and earth are shaped like spherical domes based on the Gaitian model on the universe. They would have no way to approach it since they struggle with the possibility of Beiji Xuanji being a real star.

In *Zhou Bi Suan Jing*, Xuanji is mentioned thrice, in sections 8, 9, and 12[12]:

> *What is Xuanjisiji, or the four positions of the North Star in four seasons? The first position is in the south, where the North Star is located between 23:00 and 1:00 in Summer Solstice. The second one is in the west where the North Star is located from 17:00 to 19:00 in Winter Solstice. The third is in the east from 5:00 to 7:00. See the following observation scheme.*
>
> *Xuanji is 23,000 li (1 li is 0.5 km) in diameter and 69,000 li in circumference (π was then considered to be at 3). This is a special cold and bleak area inhabitable for any living things.*

---

[10] Qian Baozong. Gai Tian Shuo Yuan Liu Kao (*Origins of Gaitian Theory*). *Collected Papers of Science History*, Vol. 1, (1958).

[11] Chen Zungui. Zhong Guo Tian Wen Xue Shi (*History of Astronomy in China*), Shanghai People's Publishing House, Vol. 1, (1980), P136–138.

[12] *Zhou Bi Suan Jing* that we discuss is based on *Interpretations of Zhou Bi Suan Jing* written by Jiang Xiaoyuan and Xie Yun, published by Liaoning Education Press in 1995.

> *Altair is xxx from Beiji (North Star)... Arithmetics tells that waiheng*
> *(the Tropic of Cancer) is 238,000 li from to Beijishu, with xuanji of*
> *11,500 li to be subtracted. ...Dongjing is xxx li from Beiji (North Star).*
> *Arithmetics shows that neiheng (equator) with xuanji of 11,500 li to be*
> *added...*

From this discussion, it can be said that Beiji (North Star), Beijishu, and Xuanji are three different concepts.

The celestial body that travels from and to the four locations was rightly believed by Chen Zunwei as being the North Star. But Zhou Bi Suan Jing clearly defines it as the then North Star, and not as Beiji Xuanji as mentioned by Chen.

Xuanji is a column between heaven and earth, the cross-section of which was the circle that the then North Star moved along, though it is still debatable as to which one was the then North Star.

Beijishu, obviously the center the circle that the then North Star moved along, is the true North Pole in the astronomical sense.

Based on the above analysis, there is no need to shy away from arguments made in the ninth and twelfth sections of *Zhou Bi Suan Jing*. They say that Xuanji, rather than an imaginative space, is situated rightly under the North Pole and above the earth's surface. Xuanji is 23,000 li in diameter, a difference of 0.77 meter in diameter from the east to the west of the North Pole as discussed in Section 8. Xuanji is called Jixia ("under the North Pole") in *Zhou Bi Suan Jing* as well.

Our understanding of Xuanji is far from complete based on what we already know. Luckily, there are some more discussions on Xuanji in the book to help us clear up some doubts, and these are given in Sections 7 and 9.

> *Jixia (A special area below the North Pole, spanning 23,000 li) is 60,000*
> *li higher than where people live. Pang Tuo Si Tui Er Xia (滂沱四颓而*
> *下, "The area becomes thicker and fatter as it goes down"). Nothing can*
> *live on Jixia. How can we know that?*

Xuanji is an entity located as high as 60,000 li above the earth's surface. The top of the entity is sharp, and it gets larger as one moves down. Xuanji's lowest part, the earth's surface, is 23,000 li in diameter (refer to

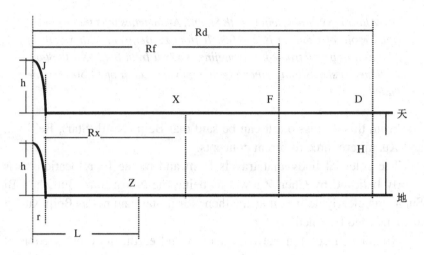

Figure 1: The shape of Gaitian Model according to *Zhou Bi Suan Jing* (side-looking semi-sectional view).

Figure 1 in this chapter). Within the area spanning 69,000 li in circumference, nothing ever lives.

I need to explain the line "Pang Tuo Si Tui Er Xia". Almost all the books that support the heaven and earth being like two spherical crowns cite this line as evidence. But they fail to note the fact that Jixia is 60,000 li higher than where people live, meaning that the heaven and earth are in no way spherical crowns. If the heaven and the earth were indeed two crowns and the center of the earth is 60,000 li higher than its border, the sentence that Jixia is 60,000 li higher than where people live would make no sense at all. This is because spherical crown means that there isn't any area on the earth 60,000 li below Jixia. Any active area suitable for human habitation can never be 60,000 li lower than Jixia. The author of Zhou Bi Suan Jing believed so as well.

In addition, if the dual spherical crowns theory is accepted, Jiaxia and the whole landmass are one, without any borders in between. This runs counter to what *Zhou Bi Suan Jing* intends to convey. As mentioned earlier, Jixia covers an area 23,000 li in diameter, which is a special cold and bleak area inhospitable for any living thing.

6. Double-layer flat universe model — The correct shape of the Gaitian Universe in *Zhou Bi Suan Jing*

Through these discussions, we know the Gaitian Model states that the heaven and the earth are in parallel and standing under the North Pole is the 60,000-li-high Xuanji which is pointed at the top and widens at the base reaching 23,000 li in diameter. We need to add three more points to this.

First, the shape of heaven on the North Pole. We know that under the North Pole stands Xuanji and that the heaven is not flat on the North Pole. The seventh section of *Zhou Bi Suan Jing* has already elaborated on this.

Jixia (a special area below the North Pole) is 60,000 li higher than where people live. Pan Tuo Si Tui Er Xia ("The area becomes thicker and fatter as it goes down"). The center of the heaven is also 60,000 li higher than the area's perimeter.

That is to say, like Xuanji, the heaven in the North Pole is also a column standing upright. The structure has been illustrated in Figure 1 of this chapter (see p.114). The picture, which is the cross-section profile of the Gaitian Model in *Zhou Bi Suan Jing*, was only half drawn, because it has rotational symmetry. The left part of the figure shows the half cross-section profile of Xuanji.

Second, the distance between the heaven and earth. If the heaven and earth were presumably in parallel, the distance between them could be easily worked out by measuring an object's shadow at a given time and using the Pythagorean proposition. According to Section 3 of *Zhou Bi Suan Jing*, if the heaven and earth were in parallel, the distance between the heaven and earth would be tantamount to that between the Sun and the earth. Then, the proposition that the heaven and earth are shaped like two spherical domes in the Gaitian Model makes no sense at all. It is because of this that Li Chunfeng lashed out at Zhou Bi Suan Jing for being "self-conflicting". In fact, the book does have another clear illustration in Section 7 of the heaven–earth distance and their supposed parallel position.

The heaven and earth could be apart as far as 80,000 li. When it is winter solstice, the heaven, though at Waiheng (exterior arm), is 20,000 li away from Jixia.

Jixia is at the top of Xuanji, 60,000 li above the ground. It is therefore 20,000 li from Heaven.

Third is the dimensions of the Gaitian Universe. The Gaitian Universe is a limited universe with the heaven and earth being two flat circles. Both circles are 810,000 li in diameter. The number was calculated based on the theory that the sun's lights reaches as far a distance as 167,000 li (refer to Note 1, Page 123 in this chapter). Please check the Sections 4 and 6 of this chapter for relevant discussions.

The daytime on winter solstice and the nighttime on summer solstice are the shortest. By observing the sunlight, we know that the universe is 810,000 li in diameter and 2.43 million li in circumference.

On winter solstice, the sunlight reaches 167,000 li out of Beiheng (North Arm). The universe is 810,000 li in diameter and 2.43 million li in circumference.

Beiheng, also known as Waiheng, is the farthest place possible from the North Pole, according to the Gaitian Model. When the Sun travels farthest from the North Pole, it is 238,000 li in radius and at the same time its light travels as far as 167,000 li. All told, the universe reaches 405,000 li in radius, and that means it has a diameter of 810,000 li.

According to *Zhou Bi Suan Jing*, all the parameters and numbers for the map are as follows:

J   North Pole (Tianzhong)
Z   Zhoudi (Luoyi)
X   Summer Solstice
F   Spring Equinox and Autumn Equinox
D   Winter Solstice
r   Radius of Xuanji in Jixia = 11,500 li
Rx Radius on Summer Solstice = 119,000 li
Rf Radius on Spring Equinox and Autumn Equinox = 178,500 li
Rd Radius on Winter Solstice = 238,000 li

L   Distance between Zhoudi and North Pole = 103,000 li

H   Distance between heaven and earth = 80,000li

h   Height of Xuanji = 60,000 li.

This is the right shape of the universe according to the Gaitian Model in *Zhou Bi Suan Jing*. The model is in line with what the book says, and the math matches as well. But why did our ancestors have the false assumption that the heaven and earth take on the shape of two spherical crowns? To better understand this, we need to take a close look at the proposition that "the heaven resembles a covering rain hat, the earth resembles a dish turned upside down".

7. Misinterpretation of the statement "the heaven resembles a covering rain hat, the earth resembles a dish turned upside down" by the ancients

That "the heaven resembles a covering rain hat, the earth resembles a dish turned upside down", as mentioned in in Section 7 of the book is an important basis for the theory of the heaven and earth taking on the shape of two spherical crowns. It is important that we dig deeper into this proposition.

The proposition in Chinese is very figurative, just like what Zhao Shuang noted:

> *The (first) appearance of anything (as a bud) is what we call a semblance (Xiang); when it has received its complete form, we call it a definite thing (Fa). The heaven is like a cover and the earth a plate. Xiang and Fa are similar in meaning and cover and plate are similar in shape. The two objects are distinguished to show one is superior to the other. Facing upward is Xiang while facing downward Fa, only different in names.*

Zhao Shuang stressed that cover and plate are mere analogies. Such a literary analogy could do no more than tell the initial shape of the universe, so it is definitely not as important as the figure that we showed earlier section. What we have discussed shows that the idea that the heaven and the earth are in parallel with each other is the foundation of the figure.

In other words, it is almost impossible to conclude that the heaven and earth take on the shape of two spherical crowns according to the proposition alone. What can be concluded is that the heaven and earth are in parallel alignment with each other. Now, let us take a close look at how the proposition is made.

Cover indicates the covering of the ancient horse carriage, that is the canopy. If we look at old paintings, we can easily see that all these covers are flat two-dimensional circles with draping around their circumferences. The center of the circle, which is connected to the umbrella handle, is protuberant. I have never seen a cover with a spherical shape.

A rain hat, which can be seen in many places, is also like a flat circle, with its center being a protuberant cone. The shape is in line with the shape of the heaven and earth shown in Figure 1, p. 114.

Plates, which were often used in ancient times, can only be flat. No one has seen a plate in the shape of a spherical crown. After all, it would then be difficult for a plate to stand on a surface without wobbling.

Generally speaking, there is no logic behind the proposition that "the heaven resembles a covering rain hat, the earth resembles a dish turned upside down". So it is implausible to prove that the heaven and earth are two spherical crowns. The reason why people believed in the proposition was because it was said by some authoritative people. But people did not know that even a wise man sometimes make mistakes.

7.1. *Wang Chong's interpretation of the Gaitian Universe*

We owe modern people's misinterpretation of the proposition to language barriers. But the ancients should not have made the same mistake. Wang Chong, a fan of debate, was a supporter of the Theory of Canopy Heavens. According to *Jin Shu Tian Wen Zhi* (*Astronomical Treatise, the Book of Jin*), he refuted the Huntian Theory based on the Theory of Canopy Heavens.

The old theory is that the heaven goes right through the earth. But if you dig deeper under the earth's surface, you will find a lot of water. So how could the heaven go through water? Definitely not. The sun travels in the heaven and it cannot go deep into the earth. The fact that the naked eye could see only as far as 10 li means it is an illusion when one sees the heaven and earth seem to converge. The sun indeed cannot go down the earth. When the sun sets in the west, people in the west actually see the sun hanging in the middle of the sky. So people think that the sun rises from near them and that it sets far from them. To make people understand it, one can walk at night, carrying a brightly lit torch. When one goes for more than 10 li, people standing in the place of one's departure cannot see the light any more. This is not because the torch is put out, but simply because it is too far to be seen. The same can be said of the sun setting in the west. The Sun and the moon are not round. They look round because they are too far away from people. It is said that the sun symbolizes fire and the moon water. Given the fact that neither fire nor water look round at all on earth, how could they become round-shaped in the heaven?

What Wang Chong said, like "the sun and the moon are not round", is also undoubtedly flawed. What is an important reference for our discussion is that Wang Chong clearly believed that the heaven and earth are in parallel alignment with each other. Such a belief was reinforced by all the analogies that he drew. What this paragraph states serves as an example of how the academics understood the Theory of Canopy Heavens in the Han Dynasty. Their interpretation of the universe was in line with conclusions that we arrived at in the previous sections. However, our forefathers who stood for the idea that the heaven and earth are like two spherical crowns did not understand the example at all.

By the way, some books cite both the theory of the canopy of heavens first and the Gaitian Theory second. The idea of a round heaven and a square earth is the first theory, while the Gaitian Theory in *Zhou Bi Suan Jing* falls under the second category. The two theories can never be seen together because the former is a mere proposition without any explanation while the latter is supported by mathematics. So it makes no sense to tell the two theories apart by placing them on the same plane.

8. Puzzles about the guidelines and origin of Huntian Universe Model

The Huntian Theory was much better received than the Gaitian Theory. The fact is that the Huntian Theory dominated the astronomy community in ancient China. But unlike *Zhou Bi Suan Jing*, which had a lot of information about Gaitian Theory, this was not systematically elaborated.

Zhang Heng Hun Yi Zhu (Notes on Armillary Sphere), quoted in *Kai Yuan Zhan Jing*, sees the Huntian Theory as a guideline. This was quoted as follows in the book:

> *The heaven and earth are like an egg. The heaven takes on a round shape while the earth is like yolk in the middle of the egg. In this sense, the heaven is big while the earth is small. Water that partly constitutes the heaven wraps around the earth like eggshell wrapping around the yolk. The heaven and the earth are composed of water and air, floating on water. The heaven travels 365.25 degrees when it completes revolving around itself for an entire circle, and it is divided right in the middle. The first 182.625 degrees part is on the ground while the other half is under the ground. Therefore, the 28 lunar mansions are half seen and half unseen. The two ends are the two poles, North Pole and South Pole. The North Pole is in the center of the heaven, which is in the True North, at 36 degrees above the ground .The part 72 degrees below the South Pole is hidden. The two poles are more than 182.5 apart from each other. The heaven revolves like a wheel incessantly, without any starting. It's spherical, and therefore is called Huntian.*

This is the basic principle governing the Huntian Theory. Though it does not provide as much information as *Zhou Bi Suan Jing*, there is indeed something we can touch upon. Let us first talk about its origin.

When the Huntian Theory was put forward is debatable. It was probably put forward between the Western Han Dynasty (206 B.C.–A.D. 24) and the Eastern Han Dynasty (25–220), but no later than 139 AD, the year Zhang Heng passed away.

Huntian Theory is believed to have been put forward in the early Western Han Dynasty because of a paragraph dedicated to the theory from *Fayan Zhongli*, a book by Yang Xiong. It goes:

> *Someone asked about the huntian theory (theory of sphere-heavens). Yangzi said: Luoxia Hong devised it, Xianyu Wangren calculated it, and Geng Zhongcheng created the armillary sphere.*

Zheng Wenguang believed that there had already existed armillary spheres and Huntian Theory during B.C. 156–B.C. 87 when Emperor Wu of the Han Dynasty was king because the armillary sphere was precisely designed based on the Huntian Theory.[13] Some academics strongly denied the existence of the armillary sphere then, but still believe it was Luo Xiahong that came up with Huntian Theory.[14] There is yet to be a consensus on this issue.

We need to note that the degree we are talking about is an ancient Chinese way of calculation. An equation has been given only for the ancient degree, as follows:

$$1 \text{ ancient degree} = 360/365.25 = 0.9856°.$$

That "the North Pole is situated 36 degrees above the earth" actually means 35.48 degrees.

North Pole's altitude above the horizon is not a constant, meaning that it changes according to latitude. But this fact was not understood by the author of the above quote, so he used the altitude as an important constant number. Because North Pole's altitude figure is precisely the same as local latitude, we can conclude that Huntian Theory was perhaps put forward at 35.48 degrees north latitude. That makes it even more complicated to trace where the theory originated. We would be puzzled if we look it up on a map for regions that had to do with the theory, including Bashu (where Luo Xiahong came from), Chang'an (where Luo Xiahong and other astronomers were called on to revise the calendar), and Luoyang (where Zhang Heng served as Taishiling, a position responsible for recording history) were very far away from 35.48 degrees north latitude. As far as I know, none of our forefathers had taken note of this fact. We may, as a result, conclude that Huntian Theory was not founded in these regions.

---

[13] Zheng Wenguang & Xu Zezong: *Cosmic History in Chinese History*, p. 69.

[14] For example, Professor Li Zhichao said in his article Studies of the Origin of armillary spheres, "any theory that supports the existence of armillary spheres before the Western Han Dynasty is improbable. Zhang Heng was the only person that came up with armillary spheres. That's why any records of spheres should be held with a grain of salt. Please refer to *Studies in the History of Natural Science*, Vol. 9, (1990).

But the question remains: where did it originate? Could the time when it was put forward be wrong too?

I raise these questions only, but will not jump to conclusions.[15]

9. Why did Huntian Universe Model go mainstream?

According to Huntian Theory, both the heaven and the earth take on the shape of a sphere, a proposition closer to modern understanding of the universe than Gaitian Theory. But Huntian Theory believes that the heaven is made of physical entities like eggshells. This means that Huntian theory is no better than Aristotle's crystal sphere theory. That said, books that lambaste crystal sphere theory shy away from swiping at problems of Huntian Theory.

The assumption that the sphere-shaped earth is floating in water is very problematic. According to this model, the moon, the sun, and stars are all attached to the inside of "celestial body". So if its lower part is immersed in water, then the sun, the moon, and stars would be in water too, which is quite counterintuitive. Later, a revised assumption that the earth was floating in air came out, which made more sense, just as what was said in *Zheng Meng-Can Liang Pian* by Zhang Zai in the Song Dynasty.

From a modern perspective, Huntian Theory is so rudimentary that it could not rival the geocentric system put forward by Ptolemy or Gaitian Theory at all. But how can such a simple theory be a mainstream school 2,000 years later?

It is very simple — because Huntian Theory describes the heaven and earth as spheres. As such, it is possible to develop a most basic sphere-based astronomy. But only sphere-based astronomy makes it possible to measure and calculate how stars run their course. Despite Gaitian Theory having its own mathematical astronomy, it does not have adequate descriptions for astronomical numbers. For example, *Zhou Bi Suan Jing* falls short of mentioning how stars move.

I said "most basic" because ancient Chinese spherical astronomy had been unable to measure up to the Greek standard. The spherical

---

[15] I have already had a preliminary, bold assumption, which is yet to be proved.

astronomy that we use today was already well developed in ancient Greece. But Huntian Theory has a fatal flaw — the earth being too big. That means astronomers or mathematicians find it impossible to develop any feasible universe geometrical model, let alone spherical astronomy.

10. Did the theory of Spherical Earth exist or not in Ancient China?

The question of whether the theory of Spherical Earth existed was not raised until it was introduced to China in late Ming Dynasty, and later accepted by some Chinese astronomers. The answer is invariably the same — yes, it did. But it is not as simple as a black-and-white question.

The following is literature proving the existence of the theory:

"The south is endless. The center of the world as I know it is the north of Yan State and the south of Yue State".

— *From Zhuang Zi Tian Xia (Chuang Tzu: the World) by Hui Shi*

The heaven and earth are like an egg. The heaven takes on a round shape while the earth is like yolk in the middle of the egg. In this sense, the heaven is big while the earth is small. Water that partly constitutes the heaven wraps around the earth like eggshell wrapping around the yolk. The heaven and earth are composed of water and air.

— *From Hun Tian Yi Tu Zhu (Legend of Armillary Spheres)*
by Zhang Heng

The heaven and earth are like a bird's egg, with the heaven wrapping around the earth like egg white around yolk. There is no end in sight when the egg starts revolving. This is the basis of Huntian Theory. It believes that half of the heaven is above the earth and the other half under the earth. The heaven, together with the sun, the moon and stars, revolves around both poles.

— *From Hun Tian Xiang Shuo (Huntian Theory) by Wang Fan*

What Hui Shi said makes sense if the earth is presumed to be round. That is why it is seen as proof of a spherical earth. The latter two quotes declare that the earth takes on the shape of a sphere.

Then can we reach the conclusion that ancient China developed the theory of spherical earth? Not really. We cannot draw such a conclusion because of what people said. In ancient Greek astronomy, the theory was closely associated with spherical astronomy. The Western version of spherical earth is characterized two features.

1.   The earth takes on the shape of a sphere.
2.   The earth is much smaller than the heaven.

It is easy to understand the first point, but not the second one. But it is important to note that rarely do we need to use the earth's size, except in such conditions as horizontal parallax and lunar eclipse. In most cases, the earth's size is necessarily and reasonably neglected. Please refer to the following:

The radius of the earth — 6371 km.

The distance between the earth and the sun — 149,597,870 km.

The proportion of the radius to the distance is 1: 23,481.

Furthermore, the earth is only the third farthest star from the sun among the nine stars in the solar system. That points to the vastness of the solar system. And there is even greater space if we take into consideration the galaxy and the alien galaxy. The earth's size therefore is negligible. It is true that the earth was the center of the universe as far as ancient Greeks were concerned, but their spherical astronomy, a science of how to measure and calculate the position of celestial bodies, could be entirely applicable to heliocenter and the modern universe system.

Back to the theory of the spherical earth in ancient China. Given that ancient Chinese people compared the heaven and earth with eggshells and yolk, it is obvious that they believed the distance between the earth and the heaven was not far. The following were their calculations about the sizes of the heaven and the earth.

The spherical heaven is 387,000 km in diameter and the earth is 193,500 li away from the core of the spherical heaven.

From Er Ya-Shi Tian (Heaven Explained)
The distance between the heaven and the earth is 678,500 li.

— *From He Luo Wei-Zhen Yao Du*

The heaven revolves 365 degrees per day and it is more than 91,000 li
from the earth's surface.

— *From Hun Tian Fu(Rhapsody on Astronomical Phenomena) by Yang*
Jiong

Take the first quote. The earth's radius is the same as its distance from the sun. Given that, it is impossible to ignore the size of the earth. Since the Ming Dynasty, academics often failed to consider this by claiming that the spherical earth theory was created in China. And many modern books also make the same mistakes.

Unfortunately, that is why ancient Chinese were unable to develop advanced astronomical ideologies, like the Greek spherical astronomy theory. Many academics blamed poor geometry and failure to use the ecliptic system as reasons why ancient Chinese astronomy failed to make concrete progress. But the fact that they believed the earth was huge in the universe could explain why it actually did not take off.

11. Yardsticks for assessing the superiority of the Universe theories

It takes a reasonable yardstick to evaluate different theories about the universe. The yardstick should also not be about whether the universe is finite or infinite, nor about whether it is idealist or materialist. History has proven that idealism is not necessarily bad and materialism is not indeed always good.

Another reasonable yardstick is to check how relevant a theory is to today's knowledge. Many researchers of science history regard the yardstick as god's truth, but actually they are not right. It takes forever for humankind to explore and understand the universe. That means even today's interpretation of the universe will never hold true forever. How Copernicus and Kepler understood the universe is not at all the truth from today's perspective, but these are ladders that lead human beings closer to truth.

Ancient astronomers saw universe models as a working hypothesis. So, the most suitable yardstick against which universe theories are assessed should be as follows.

It would be a good theory if it could take into account descriptions and predictions of the unknowns. It would be better if such descriptions and predictions finally lead to corrections to, acceptance, or denial of the theory.

What I stand for is close to falsificationism, put forward by K. R. Popper, a science philosopher, which believes that only theories that could stand the test of observations and experiments are able to advance scientific progress. Such theories are featured by falsifiability. But meaningless narratives that look always true — for example, it might or might not rain tomorrow — contribute nothing to scientific development at all.[16]

According to this gauge, theories prior to the Copernicus era are ranked as follows:

1. Ptolemy's model of universe.
2. Gaitian Model from *Zhou Bi Suan Jing*.
3. China's Huntian Theory. Xuanye Theory did not even make it to the list. The reason why Xuanye Theory did not make a difference at all is not because it is rendered groundless by observation but because it could not be falsified. No observatory evidence poses threats to the theory. That is why it contributes nothing to astronomy, and that makes the theory pointless.

The reason why Ptolemy's model of universe tops the list is that it is a geometric system that can be overturned. Since its inception, almost all the astronomical achievements in the west, including the Arabian world, were built on the system by the end of Copernicanism. In addition, it is based on the system that Tycho de Brahe's system, Copernicanism, and Kepler's theory came into being.

---

[16] Popper's theory is specifically discussed in his book *Conjectures and Refutations* (1969) and *Objective Knowledge* (1972). Both have Chinese translations published in 1986 and 1987 respectively by Shanghai Translation Publishing House. There had been a lot of developments in science and philosophy after Popper. The most recommended book is *What Exactly is Science* by A. F. Chalmers, published by the Commercial Press in 1982.

I have managed to prove that Gaitian Theory can also be categorized as a reasonable geometric system, though it is rather rudimentary.[17] Its universe model has a clear geometric structure, supported by specific numbers. People drew the conclusion that "the distance of the sun's shadow differs by 1.31 cun (1 cun = 1/3 decimetre) every 1,000 li (1 li = 0.5 kilometre) from north to south". It is a proposition inferred from a self-evident truth that "the heaven and earth are a plane in parallel". Furthermore, the conclusion can be falsified. In 724, Yi Xing and Nan Gong organized a campaign that aimed to measure the distance of the sun's shadow in different parts of China. It proved the proposition wrong.[18] Then people declared Gaitian Theory groundless. The reason why Gaitian Theory ranks before Huntian Theory is that it is a really valuable attempt to be accepted as truth.

Though Huntian Theory failed to develop into a decent theory, it could make predictions and descriptions of unknown astronomic phenomenon, thus making possible the thousands-of-years of development of traditional Chinese astronomy. Huntian's propositions, say, that the earth is like a sphere, could also be overthrown because they go against facts. Gaitian's proposition that the heaven and earth are flat planes in parallel could also be overturned.

Though lagging behind other countries in the universe modeling system, China achieved a lot in mathematical astronomy, something that has eluded many Westerners. This is because Chinese people valued practical use, so they put on hold the development of any theoretical knowledge, including the universe model. Ancient Chinese astronomers used algebra and equations to describe how celestial bodies moved, and with good results, using a method employed by ancient Babylonian astronomers as well. They simply paid no attention to what the universe structure was like. Ancient Chinese did not associate the universe model with mathematical astronomy. This is different from the west, which deducted mathematical astronomy directly from the universe's geometrical model.

---

[17] For more details, please refer to Jiang Xiaoyuan's *Zhou Bi Suan Jing* — Ancient China's only axiomatic attempt, *Journal of Dialectics of Nature*, 1996, Vol. 18 (3).

[18] Distance of the sun's shadow also differs as latitude changes. And the difference is even more wild, 1.31 cun every 200 li. Please refer to page 164 of *History of Astronomy in China*, compiled by a research group for astronomic history.

12. Interactions between the Heaven and Mankind

The earlier sections explained physical properties from today's point of view. Finally, we need to discuss the concept that the heaven and humans are highly interactive, an important aspect of the universe concept. This section makes general and supplementary comments only, given that plenty has been discussed in my book *Tianxue Zhenyuan (The Truth of the Sciences of the Heaven).*[19]

In narratives of Chinese ancients, Tian (heaven) represented the nature or the universe. The nature they referred to was not the nature we know of in modern science, not material, nor had it anything to do with human understanding and will. Instead, it was moral, emotional, and willed giant creatures with people included. In this sense, ancients used Yuzhou (universe) rather than Tian (heaven) as their choice of word.

The heaven and man are interactive as people doing something good will move the heaven while those who do bad things will be damned by the heaven. On the other hand, the heaven could nourish everything with its compassion and warn people with its rage. The dynamics between the heaven (tian) and human (ren) are what is known as uniformity between man and heaven.

With such an outlook of the universe, the Western idea that human-kind can change and conquer the nature did not hold sway at all. Ancient Chinese believed that natural resources are the gift of the heaven. Solar eclipses, comets, droughts, and earthquakes were seen as the heaven's warning about lapses in imperial rule. To ease the heaven's anger, emperors must pray and apologize to the god, otherwise they would be stripped of their ruling power by the heaven. That is why any new emperor would invariably declare that he was mandated by the heaven.

In addition, the so-called "governing based on four seasons" really caught on in ancient China, which is exemplified in Volume 11 Yin Yang Yi, *Chun Qiu Fan Lu (Luxuriant Dew of the Spring and Autumn Annals)*, by Dong Zhongshu in the Han Dynasty.

> *Like human beings, the heaven is sometimes happy, but sometimes sad, ill-tempered. Given that both are so like each other, we can say man and heaven are unified. Spring sees everything grow given that the season*

---

[19] Please refer to second and forth chapters in *Tian Xue Zhen Yuan (The Truth of the Sciences of the Heaven).*

contains celebratory mood. Summer nourishes everything because it is
a merry season. Winter, filled with grief, sees nothing come out. Working
in sync with the heaven helped emperors govern the country, and vice
versa. That's why the way emperors governed should be the same as the
way the heaven governs nature.

Emperors were not alone. Prime ministers were also running their
business in accordance with four seasons. This was said in *Shih Chi-Chen*
Cheng Xiang Shi Jia (The Historical Records — Hereditary House of
Prime Minister Chen):

The prime minister is supposed to assist the emperor in doing what should
be done in different seasons. For example, what kinds of decrees should be
issued in what season so that everything grows in sync with seasons.

Till today, "timeliness" is still used as a legacy. As far as ancient
Chinese are concerned, if decrees were not issued at the right time, the
climate would turn to be pretty ugly and floods, droughts, insect plagues
ensued. All these extreme weather events were interpreted as the heaven
expressing its rage.

13. Can we understand the Universe?

Unity between heaven and humankind, of course, runs counter to what
modern science is all about.[20] But such a conception echoes modern cosmology as it also acknowledges its limited understanding of the universe.
For example, after saying that the universe is double-layer circular planes,
Zhou Bi Suan Jing also claims that:

Beyond what we are capable of observing, nobody knows what's out
there and whether people are ever going to understand that. It is corre-
spondent to the Confucian's proposition that you are a wise man if you
know that you don't know.

Given that ancient Chinese knew they had limited knowledge, they
were much better than some modern Chinese who make arbitrary find-

---

[20] But some people crown the concept as "organic view of nature" and they associate it
with concepts concerning modern environmental protection. Such an association is far-
fetched and seems to overstate ancients' achievements.

ings, are paranoid, and are mere herd-followers. Zhang Heng said in *Ling Xian*,

> *I know nothing more than this. What I don't understand is the universe.*

Similar to this is Yang Shen in Ming Dynasty, who said:

> *You will not see things clearly unless you are not within them. Given that people live in the heaven and earth, how could they truly understand what the heaven and earth are.*[21]

He means that humankind is part of the universe, and so they are not capable of truly understanding the universe. Compared with the ancients, modern cosmologists have a deeper and more profound understanding of human limitations. One of them is J. A. Wheeler, who stated in a speech an imaginary conversation between people and the universe:

> *The universe: I am a giant machine. I supply the space and time for your existence. There was no before before I came into being, and there will be no after after I cease to exist. You are an unimportant bit of matter located in an unimportant galaxy.*
>
> *People: Yes, oh Almighty Universe, without you I would not have been able to come into being. Yet you, great system, is made of phenomena; and every phenomenon rests on an act of observation. You could never even exist without elementary acts of registration such as mine.*[22]

Wheeler tried to say that without the universe, there would be no human understanding, and without people, there would be room for the existence of the universe. The universe he meant is well beyond the objective universe.

---

[21]Yang Shen. *Shen An Quan Ji (Shenan Literary Collections)*. Vol. 74. "Song Ru's comments on the heaven".

[22]Fang Lizhi. *Hui Le Yan Jiang Ji: Wu Li Xue He Zhi Pu Xing (Wheeler's Speech Collection — Physics and Plainness)*. Anhui: Science and Technology Press, (1982), p. 18.

Chapter 8

Sino-foreign Exchanges in Ancient Tianxue (1)

1. Sino-foreign exchanges and honor of the ancients

As has been noted, the history of ancient Chinese tianxue is the history of Sino-foreign exchanges in astronomy. Available evidence of different kinds suggests that the scale and frequency of astronomic communication between ancient China and foreign countries are probably way beyond today's common expectation. However, to explain this, first we have to sort out the problem about the honor of our ancestors.

What for? Actually, this was not my intention. But so far, of all the academic works by scholars from home and abroad on historical Sino-foreign exchanges, over 90% have dwelled on the spread of Western astronomy in China, with only a few studying the aspect the other way around.[1] Hence, some complained indignantly: You always study Western

---

[1] Such studies can certainly be found in some other fields. For example, in Joseph Needham's masterpiece, the westward spread of gunpowder, papermaking and printing received pleasingly detailed description. In terms of the field of astronomy alone, a few relevant studies can also be located if you try hard, like Han Qi and Duan Yibing's "French Sinologist Edouard-Constant Biot's Study on Ancient Chinese Astronomical Records and His Contributions to Western Astronomy" published on the China Historical Materials of Science and Technology, Vol. 18, No. 1 (1997). However, that paper revolves around Biot's study of the historical Chinese astronomical texts translated into Western languages by Jesuit missionaries in the 17th and 18th century, which is quite different from the study of

influence in China, attributing this and that glorious achievement of our ancestors to the West; aren't you discrediting them? Aren't you tarring their honor? Why not focus more on Chinese influence in the West?

Their indignation is not unreasonable. All those past achievements that we have taken pride in and regarded as "the quintessence of Chinese culture" were suddenly proved to have foreign roots or come as a result of foreign impact — this gives some a sense of loss. However, what is happening in reality is more reasonable than their indignation.

First, academy upholds the pursuit of truth; the objective facts ill not change themselves for subjective human emotions. If the abundance of domestic studies on "west-to-east spread" is due to limited research materials or language barriers, then many foreign scholars (quite a few are ethnic Chinese) are free of those constraints; why both Chinese and Westerners, in China or other countries, published academic reports that mostly centered on "west-to-east spread"? I am afraid it can only demonstrate that: in the historical materials, "west-to-east spread" does dominate. If this is indeed the truth, no expression of indignation can change it.

Second, we must forsake the old-fashioned belief that it is always an honor for Chinese ancestors to spread something of their own to others and a shame to receive something that's imported. In today's China after the reform and opening up, are we all not taking pride in the introduction of foreign high-techs? Are we all not taking pride in catching up with the developed countries on advanced theories? Are we all not taking pride in being integrated into the world? Why should we not apply the same standard to our ancestors?

Actually, in the history of Sino-foreign astronomic exchanges, there are evidences that Chinese ancestors have adopted Western knowledge. This shows that opening up is not a national policy that was first developed in the 1980s, but has always been a brilliant Chinese heritage through the thousands of years. The Chinese nation has always been tolerant and broad-minded, never saying no to new theories and knowledge

---

the historical spread of Chinese astronomy and can also hardly rival such studies as focus on the westward spread of papermaking. Nevertheless, in the 15th section of this chapter, we can see an example that might truly show the introduction of Chinese astronomy into the West.

from abroad. Self-isolation and ignorant arrogance are only bucking the historical trend.

So to sum up: "always studying Western influence in China, attributing this and that glorious achievement of our ancestors to the West" (not really, in fact) actually burnishes, not tars, the honor of our ancestors.[2]

2. A bird's-eye view: The inextricable link between astrology and religion

A rough bird's-eye view of the historical Sino-foreign exchanges in astronomy reveals that there's an extricable link between astronomical exchanges and religion.

The spread of Western astronomy in China saw a peak during the period between the Six Dynasties and the Tang Dynasty, when Western knowledge in astronomy made inroads into the Chinese mainland together with Buddhist philosophy through the major medium of India. Some of the most meaningful and relevant texts, like *Qi Yao Rang Zai Jue*, are exactly preserved in the form of sutras.

The period between the Yuan Dynasty and the early Ming is regarded as another peak of Sino-foreign astronomic exchanges. This time, thanks to the medium of Islamic astronomy, Western astronomy once again made its way into China with the rise of the Mongol Empire that straddled Eurasia. This second peak has a lesser religious tone and has begun to see some traces of "east-to-west spread" (we will talk about an example later).

The third peak appeared in the late Ming. Now Western astronomy drew on no mediums, and poured into China directly. Though at the time it (and other knowledge of Western natural sciences) mainly served as a tool for Jesuit missionaries to preach in China and thus took a strong religious hue, it generated the most outstanding effect — ancient China was almost led to the gate of modern astronomy. Unluckily, mainly due to Chinese society's own problems, we stopped there without entering. After

---

[2] For the record, this conclusion is first heard from my student Dr. Niu Weixing. I dare not claim credit for it.

three hundred years, we finally crossed the threshold, but by then we had already lagged far behind the West.

Of the said three peaks, the first has been elaborated on in my book *Tianxue Zhenyuan*,[3] but still this chapter will offer some very interesting new findings that my students and myself have found in the recent years. To the second peak, our predecessors have already devoted pages, yet without systematic and comprehensive description.[4] In this chapter, I will also present some new explorative attempts of mine during these years. As regards the third peak, it is the main topic of the following chapters.

However, we should never be confined by the given three peaks (which has been quite common) and think that there is not much else about the historical Sino-foreign exchanges in astronomy. As a matter of fact, in the past thousands of years, bilateral astronomic communication has never stopped. The early period in particular deserves attention, for what we know today is probably only the tip of the iceberg. Many shocking facts remain to be disclosed, and many shocking secrets are yet to be uncovered — this is exactly the glamour of the history of Sino-foreign astronomic exchanges.

3. Shocking similarity between the universe model of *Zhou Bi Suan Jing* and that of ancient India

According to convincing conclusions drawn by modern scholars, *Zhou Bi Suan Jing* was completed around 100 B.C. Throughout history, it has been considered, without doubt, as one of the purest national treasures. To discuss whether the book is mixed with foreign elements seems to be whimsical. However, if we first direct our eye from ancient Chinese astrology to the astrological progress of other ancient civilizations in the world before poring over the original text of *Zhou Bi Suan Jing*, we will find, much to our surprise, that the idea is not only not so whimsical but of profound significance in the history and philosophy of science.

---

[3] Refer to Jiang Xiaoyuan: *Tianxue Zhenyuan*, Chapter 6.
[4] The *History of Chinese Astronomy* edited by Chinese Astronomy History Research Group has relevant description which is sadly scattered in different chapters. Later, there has also been at times several theses on the topic.

From the introduction given in Chapter Seven (including a specific account from the original text of *Zhou Bi Suan Jing* and my intimate examination of several key problems), we already know that the universe model of canopy-heavens in *Zhou Bi Suan Jing* has the following features:

1. The sky and the earth are two parallel discs 40,000 km apart.
2. From the center of the earth rises a giant column (namely, the 30,000 km-high "xuanji", whose base is 11,500 km in diameter).
3. The creator of this model of universe has identified his location on the earth disc, which is not around the center but in the southern part.
4. The other end of the column connects to the sky, with their junction forming the North Pole.
5. The sun, the moon, and the stars revolve around the North Pole in the sky.
6. The sun has multiple concentric orbits, and it migrates between the orbits periodically every half year (completing one cycle in a year).
7. Sunlight radiates all around, and the radius limit is 83,500 km.[5]
8. The aforesaid movement of the sun can, to an extent, explain the day/night cycle and some astronomical phenomena that occur during the apparent motion of the sun.
9. In all the calculations involved, Pi is rounded to 3.

To my great surprise, I find that the nine features were also seen in the ancient Indian model universe, without exception! Such a coincidence is by no means a pure accident and deserves attention and investigation. In what follows, I will first present the result of a preliminary comparison — more in-depth research might have to wait until some other day.

Records about the ancient Indian cosmological model are mainly seen in the Puranas. The Puranas, held holy by the Hindus, include ancient historical texts and are encyclopedic. Though it is hard to tell

---

[5] "Sunlight travels 83,500 km" is one of the axioms accepted by *Zhou Bi Suan Jing*. Those axioms form the premise for all calculations in *Zhou Bi Suan Jing*. For more details, refer to Jiang Xiaoyuan: "*Zhou Bi Suan Jing* — the Only Attempt of Ancient China at An Axiomatic System", the 7th International Conference on Chinese Science History (Shenzhen, China, Jan. 1996), published in *Journal of Dialectics of Nature*, Vol. 18, No. 3 (1996).

when they were composed, the cosmology contained in it, the scholars believe, can be traced to the Vedic period — before 1,000 B.C. or so, and thus is very ancient. The cosmological system delineated by the Puranas can be summarized as follows:

> The earth is like a round plate, at whose center stands a towering mountain called Meru (or Sumeru). Mount Meru is surrounded by a ring of land which is surrounded by a ring of ocean — seven alternate rings of land and ocean in total surround the mountain.
> India is located to the south of Mount Meru.

In the sky parallel to the earth are a series of celestial wheels whose common axis is Mount Meru; right above Mount Meru lies the North Star (Dhruva), around which all other celestial bodies — including the sun, the moon, other fixed stars and the Five Planets (Mercury, Venus, Mars, Jupiter and Saturn respectively) — revolve, carried along by the celestial wheels. The existence of Mount Meru can explain the alternation of day and night. The celestial wheel that carries the sun has 180 orbits; the sun switches to a different orbit every day, and repeats the process in reverse after half a year — this leads to the variation in the azimuth of sunrise throughout the year...[6]

Furthermore, the *Shi Jia Fang Zhi (Sakya Chronicles)* written by Tang Dynasty author Shi Daoxuan also recorded the universe model of ancient India, adding more details to the information given above:

> *...Mount Meru, or Sumeru as it's called in the sutras, lies in a sea, outside the Golden Wheel. Its upper half stands over eighty thousand yojanas above sea level. The sun and the moon circle its waist. Around Mount Meru are seven rings of golden mountains and between each two of them seawater runs, carrying with it eight virtues.*

Then the Chinese version of the Buddhist text *Li Shi Apitan Lun* (Taisho Tripitaka No. 1644) has mentioned in its fifth volume the limit

---

[6] D. Pingree. *History of Mathematical Astronomy in India*, collected in *Dictionary of Scientific Biography*, Vol. 16, New York, (1981), p. 554. It's actually a monograph on the mathematical astronomy of ancient India, not a biography.

sunlight can travel in distance and, based on it, explained the apparent motion of the sun:

> *The radius of sunray is 700,021,200 yojanas, and the circumference is 2,100,063,600 yojanas. When the sun rises from the southern dvipa ("continent") of Jambudvipa, it is dusk in the northern dvipa of Uttarakuru, noon in the eastern dvipa of Furvavideha, and right at midnight in the Western dvipa of Aparagodaniya. So the day cycles through different periods because of the sun.*

From this paragraph and substantial astronomical data from Buddhist texts, it can be seen that the value of Pi adopted is precisely 3.

Based on these records, it can be seen that the ancient Indian universe model is astonishingly similar to the canopy-heavens universe model in *Zhou Bi Suan Jing*, almost identical in all the details. A few of these are listed in the following:

1. In both models, the earth and the sky are parallel discs.
2. Both "Xuanji" and "Mount Meru" play the role of a "heavenly column" at the center of the earth.
3. Both the State of Zhou and India are placed on the southern half of the landmass in their respective models.
4. The axis of rotation for celestial bodies — the North Pole — is positioned right above both "Xuanji" and "Mount Meru".
5. The sun, the moon, the stars, and the planets in the sky revolve on a circular plane around the North Pole.
6. If the "seven rings of land and ocean" outside Mount Meru conjures up the digits in the "seven Rings and six Intervals" in *Zhou Bi Suan Jing*, then the 180 orbits of the solar wheel in the Indian universe are in perfect congruity with the 7 Rings and 6 Intervals (the sun is also continually travelling between the 7 Rings).
7. It is worth mentioning that in *Zhou Bi Suan Jing*, the distance between the sky and the earth is 80,000 li (40,000 km); rather coincidentally, Mount Meru also stands as high as "80,000 yojanas" above the sea level, with sky wheels above its peak. In both models, the distance between the sky and the earth is measured at 80,000 units. Could it be by sheer chance?

8. In both models, the sunlight radiates within a limited distance, which helps illustrate the rise and fall of the sun, the cycling of four seasons, the variation in the length of day and night, and other astronomical phenomena related to the apparent solar motion.
9. In both models, the circumference ratio Pi (π) is rounded to 3.

In the history of human civilization, spontaneous development of cultural diversity is entirely reasonable. Therefore, many civilizations have similarities that may be just happenstance. However, the universe model in *Zhou Bi Suan Jing* and that in ancient India are similar to a great extent — from the general structure to many details, the consistence is prevalent. If we are to explain this with "coincidence," it will seem too farfetched.

4. Astonishing knowledge about the climate zones

Zhou Bi Suan Jing had information about the climate zones that will be familiar to modern people. This is astonishing — because in the past 2,000 years, the traditional Chinese astronomy has lacked and rejected such knowledge.

That knowledge is mainly manifested in section 9 of *Zhou Bi Suan Jing*:

> *Under the North Pole, nothing grows. How do we know? ... Around the North Pole, there's perennial ice that do not melt even in summer.*
>
> *The Middle Ring is 75,500 li from the Zhou kingdom. Around the Middle Ring, there are grasses living even in winter, and they thrive in summer. There yang is strong and yin is weak, so living beings do not die, and crops ripen twice a year.*
>
> *Around the North Pole, some plants grow in the morning and die in the evening; they are winter plants.*

Here, some explanations in advance are needed:

In the second paragraph cited above, the so-called "Middle Ring" is the zone located "outside the Inner Ring, inside the Outer Ring" as pointed out by Zhao Shuang in his notes. This zone corresponds exactly to the tropic zone on earth (between 23°30′ south latitude and 23°30′ north latitude), although the concept of Earth was not established in *Zhou Bi Suan Jing*.

The third paragraph cited above says that around the North Pole, "some plants grow in the morning and die in the evening". Here we have to consider the ability of the *Zhou Bi Suan Jing* to make deductions and descriptions about the polar day and polar night. According to the introduction in the former chapter, the bottom diameter of "xuanji" in the center of the earth disc is 23,000 li, meaning that the radius is 11,500 li; the sunshine radiation limit set by *Zhou Bi Suan Jing* is 167,000 li. Thus, as can be seen clearly from Fig. 1 in Chapter 7, from the Spring Equinox to Autumnal Equinox each year, around "xuanji" polar day appears — with round-the-clock sunlight, and from the Autumnal Equinox to Spring Equinox, polar night appears — sunlight cannot reach within "xuanji" during this period. This is interpreted by Zhao Shuang in his annotation as: "Under the North Pole, it is day from the Spring Equinox to Autumnal Equinox, and night from the Autumnal Equinox to Spring Equinox," because day and night each occupies half a year.

The correctness of the knowledge about climate zones in *Zhou Bi Suan Jing* is undisputed. But this knowledge has not been a constituent of the traditional Chinese astrology in the past 2,000 years. And this can be discussed from several aspects.

First, Zhao Shuang, who annotated *Zhou Bi Suan Jing*, surprisingly, impugned this knowledge. For instance, when making notes for the sentence "Around the North Pole, there's perennial ice that do not melt even in summer", he wrote: "Ice does not melt. From this, we can infer that on the Summer Solstice, under the Outer Ring it is winter when all things die — this uses distance from the sun to explain the appearance of winter and summer, instead of the yin-yang theory. I doubt it". Moreover, concerning the tropical zone, in which "There are grasses living even in winter", "Yang is strong and Yin is weak", and "crops ripen twice a year", Zhao wrote: "This proposes that between the Inner Ring and the Outer Ring, summer stays forever. But eternal summer in such a vast area, I have never heard of it". He had never heard of that. Judging from the notes that Zhao Shuang made for the whole *Zhou Bi Suan Jing*, we are sure that he was a qualified astronomer during his times, but how come he had never known of the climate zones? The only reasonable explanation seems to be that this knowledge was not part of the traditional Chinese system of astronomy, so for the majority of Chinese astronomers then, it was alien, out of tune with their entrenched background knowledge and thus hard to give credence to.

Second, the theory of sphere-heavens, the mainstream theory in ancient Chinese astronomy, did not have a correct conception of Earth, so it could not have raised any question about the climate zones.[7] That was also exactly why in the late Ming Dynasty when Jesuit missionaries came to China and introduced the five climate zones in their works to Chinese readers, the theory was still regarded as a novelty that was as yet unheard of.[8] But it was these works in Chinese that made Chinese scholars accept the climatic zonation of earth. However, in the early Qing marked by the ethos of "Western learning originated from China", people like Mei Wending sought to find a Chinese root for the climatic zonation theory, and they found *Zhou Bi Suan Jing* — they thought that the Greeks, Romans and Arabs would not have had the astronomical wisdom without Chinese books like *Zhou Bi Suan Jing* that were spread to the West in the ancient times.[9]

Now we are confronted with a series of tough questions.

Since the theory of sphere-heavens did not have a correct conception of Earth and could not have broached the climate zones, then why did *Zhou Bi Suan Jing*, lacking the same conception, record this knowledge?

Assume the author of *Zhou Bi Suan Jing* lived in the north temperate zone, and he, relying on his experience that the temperature tends to dip

---

[7] Bo Shuren. "Concerning the *Gai Tian Shuo* (A Theory of Hemispherical Dome) in *Zhou Bi Suan Jing* (Arithmetical Classic of the Gnomon and Circular Paths)", published in *Studies in the History of Natural Sciences*, Vol. 8, No. 4 (1989).

[8] Among the earliest of such works is the *Apología de la Verdadera Religión* published in 1593; the most influential one is Matteo Ricci's *Great Universal Geographic Map* published in 1602; and in 1623, Jules Aleni composed *Zhi Fang Wai Ji*, which gave a more thorough description.

[9] For more details, refer to Jiang Xiaoyuan. "A Tentative Discussion on 'the Theory of Western Learning being of Chinese Origin' of the Qing Dynasty", *Studies in the History of Natural Sciences*, Vol. 7, No. 2 (1988).

farther north and climb farther south, could derive the conclusion that in the North Pole "there's perennial ice that do not melt even in summer", and in the tropic zone "crops ripen twice a year", then why could not the advocators of sphere-heavens?

Furthermore, Zhao Shuang, who annotated *Zhou Bi Suan Jing*, should be a believer in the theory of canopy-heavens. But how come even he questioned that knowledge?

Thus, it seems necessary to consider the possibility of its foreign origin.

Knowledge of the earth as a globe, the geographic longitudes and latitudes, the climate zones, and the like have been developed to maturity by ancient Greek astronomers, and it has been used till date. The climate zones have been described in the works of Aristotle, and took full shape later in *Geography* written by Eratosthenes (275–195 B.C.), "father of geography" — the tropical zone being between 24° S and 24° N, the south and north frigid zones from the 66° latitudes to the geographical poles, and the south and north temperate zones between 24° S–66° S and 24° N–66° N, respectively. In terms of the time, ancient Greek astronomers established this knowledge long before the book *Zhou Bi Suan Jing* was completed. Could it be that the author of *Zhou Bi Suan Jing* directly or indirectly acquired this knowledge from the ancient Greeks? This is indeed a thought-provoking question.

5. Ecliptic coordinates

The traditional Chinese astronomy based on the theory of sphere-heavens fully adopted the equatorial coordinate system. To set up such a system, first it is necessary to identify the "beiji chudi" — or the geographical latitude as called in modern times — of the observing point. The coordinate system mapping the celestial sphere is formed by the twenty-eight Lunar Mansions; in this system, the measurements of "ruxiudu" and "qujidu" are equivalent, respectively, to the modern concepts of right ascension and declination, in terms of both nature and function. Matching the equatorial coordinate system are all the equatorial angle-measuring instruments of ancient China, represented by the armillary spheres. The equatorial feature of traditional Chinese astronomy has also drawn special attention from modern Western scholars, because Western astronomy, since the times of ancient Babylon and Greece, had stuck to the ecliptic coordinate system for over two thousand years until the late 16th century when Europe finally witnessed the advent of a monumental equatorial astronomical instrument, which is regarded as a great invention of Danish astronomer, Tycho Brahe. Thus, in the study of foreign and domestic scholars of modern times, the equatorial feature of traditional Chinese astronomy is already an accepted fact.

However, there is no manifestation of that feature throughout *Zhou Bi Suan Jing*.

To start with, in the book, the twenty-eight Lunar Mansions are iden- tified as being distributed along the ecliptic. This is made clear in both the text and the annotations of Zhao Shuang. For example, in Section 4, Volume 1 of *Zhou Bi Suan Jing*, it says:

> *The moon orbits frequently near the twenty-eight Lunar Mansions, and the sun also never travels far from them.*

Here Zhao annotated that:

> *Draw a circle south of the Inner Ring and north of the Outer Ring — this is exactly the ecliptic and the twenty-eight Lunar Mansions scatter along it. The moon's orbit is not parallel to the ecliptic; every $5\frac{20}{23}$ months, it overlaps the ecliptic, a phenomenon called "heshuo jiaohui" in Chinese and related to the prediction of lunar eclipses, so the moon is described as orbiting along the twenty-eight Lunar Mansions belt (or "yuan xiu" in Chinese). The sun orbits on the plane of the ecliptic. Its position is identified with reference to the mansions, namely, "xiu zheng".*

Based on the materials given, it can be inferred that "the ecliptic" mentioned above is indeed identical with that of modern astronomy — the ecliptic is, in the first place, defined as the apparent annual path of the sun. Moreover, in his note for the "Figure of Seven Rings" in the 6th section of *Zhou Bi Suan Jing*, Zhao mentioned explicitly for a second time that:

> *The yellow one represents the ecliptic, along which are distributed the twenty-eight Lunar Mansions and on whose plane the sun, the moon and the five planets also orbit.*

Of course, it is on the plane of the ecliptic that the sun and the moon orbit (though technically the plane of the lunar orbit is inclined to the ecliptic by about 5°, but in ancient literature this is often ignored).

Then, *Zhou Bi Suan Jing* tried to pinpoint the reference stars of the twenty-eight Lunar Mansions in a horizontal coordinate system. This is detailed in Section 9, Volume 2 of the book. Since such a system uses the observer's local horizon as the fundamental plane, the value of the coordinates changes with latitude. Consequently, the horizontal system cannot be applied to the star maps that locate astronomical objects. Yet, it is exactly such a map that the book's attempt to measure the distance in degrees between the reference stars, namely *judu*, produced. So, judging from the basic knowledge of modern astronomy, its plan to locate the stars is a failure. Besides, what also deserves attention is that the only *judu* value the book offered was 8°, the judu of the Ox Mansion reference star, which is found to be a figure derived from calculation based on an equatorial coordinate system (the value should be 6° if adopting the book's horizontal system).[10]

Dealing with the coordinates on the celestial sphere, *Zhou Bi Suan Jing* does possess great flaws: on the one hand, it clearly believes that the twenty-eight Lunar Mansions gather along the ecliptic; yet on the other hand, it attempts to determine their *judu* in a horizontal coordinate system, and even then the only figure it provides as an example was proved to be from an equatorial system. This is worth some deep thinking, as it may offer some important clues.

5.1. *Pseudo-Ecliptic analysis in ancient China*

By the way, here we might have some more discussion on the ecliptic coordinates. Though traditional Chinese astronomy adheres to the equatorial coordinate system, it is not oblivious of the ecliptic. After all, the ecliptic, on whose plane the sun and the moon orbit, is unlikely to be ignored by anyone who has accumulated a certain amount of

---

[10] Bo Shuren. "Concerning the Gaitian Shuo (A Theory of Hemispherical Dome) in *Zhou Bi Suan Jing* (*Arithmetical Classic of the Gnomon and Circular Paths*)", published in *Studies in the History of Natural Sciences*, Vol. 8, No. 4 (1989).

astronomical knowledge. But the ancient Chinese had kept a system of ecliptic coordinates that was different from that of the West and seen by modern scholars as based on a "pseudo-ecliptic". The pseudo-ecliptic has proposed a well-founded ecliptic plane, but it has never defined the ecliptic pole. It adopts the point of intersection where the right ascension line stretching from the north celestial pole to the south meets the ecliptic plane to measure the position of celestial objects, with a result that greatly varies from their true ecliptic longitudes and latitudes. This serves as a telling example of the backwardness of ancient China in geometry.

6. Secrets in the Sino-foreign exchanges behind *Zhou Bi Suan Jing*

Repeated perusal of *Zhou Bi Suan Jing* gives the impression that its author, while versed in traditional Chinese astronomy, has also acquired from somewhere else some new methodologies — of which the most important one is the axiomatic method of ancient Greece (the book marks the only serious attempt of ancient China at the axiomatic method) — and some new knowledge, like the Indian theory of cosmic structure and the very Greek wisdom about the five climatic zones. These alien methodologies and knowledge obtained by unknown means are misfits in the traditional Chinese system of astronomy, but apparently the author thought highly of them and thus strived to integrate them in pursuit of a new type of astronomy that mixes Chinese and Western elements. His effort was successful to a certain extent. *Zhou Bi Suan Jing* indeed developed a unique system of its own, though it inevitably has some flaws.

So, who on earth is the author of *Zhou Bi Suan Jing*? What special encounters did he have during his composition of the book? Where exactly do those foreign elements come from?... All these questions remain unanswered today. Compared with the three peaks of Sino-foreign astronomical exchanges in later times, the book's involvement with Indian and Greek astronomy is more special and curious. Perhaps this is exactly what the tip of an iceberg looks like. I feel it keenly that behind *Zhou Bi Suan Jing*, there lies a great secret about ancient Sino-foreign cultural exchange.

7. Changchun Hall Symposium moderated by Emperor Wu of Liang

However, *Zhou Bi Suan Jing*'s story with Indian astronomy has not ended there. Entering the reign of Xiao Yan, Emperor Wu of Liang, a new episode took place — the famous Changchun Hall Conference.

The Changchun Hall Conference in which Emperor Wu gathered his court for academic discussion is a very notable event in the Chinese cultural history. It is recorded in the *Astronomical Treatise, the Book of Sui (1)*:

> *Emperor Wu of Liang convened a meeting in the Changchun Hall to discuss cosmology. He expressed views exactly the same as Zhou Bi Suan Jing. In fact, it's his attempt to oust the theory of sphere-heavens with a different set of ideas.*

Modern scholars often pay little attention to the Conference. Science historians, who come to focus on it because it touches upon theories of the universe, also tend to be misled because the record above slurs over the topic with a rough summary that "his views are exactly the same as *Zhou Bi Suan Jing*". An exception is Chen Yinke, who noted the event from the perspective of Sino-foreign cultural exchange and said:

> *(The theory expressed by Emperor Wu of Liang) is clearly from India. And this theory must have superiority over the theory of sphere-heavens, since the emperor tried to banish the latter with the former. Then as Astronomical Treatise, the Book of Sui has said that the theory of Emperor Wu is identical with the theory of canopy-heavens, it must also be a brand-new type of canopy-heavens theory that is imported from India. When Ko Qianzhi and Yin Shao (both are famous scholars) were learning the mathematical astronomy in Zhou Bi Suan Jing from learned masters including Cheng Gongxing, Tanying and Famu, they were, in effect, learning from Buddhism its new India-imported theory of canopy-heavens. So Ko Qianzhi, using the traditional way of calculation, worked on the motion of the sun, the moon and the five planets for years with no breakthroughs until Buddhists introduced from India the new mathematical astronomy.*[11]

---

[11] Chen Yinke. "Cui Hao and Ko Qianzhi", collected in *Jinmingguan Conggao Chubian*, Shanghai Classics Publishing House, (1980), p. 118.

Chen's opinion has a point — he connected Emperor Wu's canopy-heavens theory to Buddhism and Indian astronomy. Nevertheless, he is being too arbitrary by concluding that "this theory must have superiority over the theory of sphere-heavens, since the emperor tried to banish the latter with the former". But of course, Chen is not an expert in the history of astronomy, so we should not be too hard on him.

Another scholar from Japan named Yamada Keiji proposed that the passage "Tianti Zonghun" in *Kai Yuan Zhan Jing* was exactly the same content as Emperor Wu's speech in the Changchun Hall Conference.

Moreover, he also said that the Conference took place before the construction of the Tongtai Temple,[12] though without providing any evidence. Actually, the time sequence does not really matter, but he linked the universe model proposed by Emperor Wu in his Changchun Hall speech to the construction of the Tongtai Temple, and this is of great significance (detailed below).

Now let us look at the opinion of a domestic science historian:

Compared with the theory of canopy-heavens, the theory of sphere-heavens is a huge step forward. However, in the history of science, there's always someone trying to put the clock back. When in around 525 A.D., Emperor Wu, a faithful Buddhist disciple, gathered a group of people in the Changchun Hall to discuss cosmic theories, much surprisingly, they, including the emperor himself, all opposed the theory of sphere-heavens in favor of the theory of canopy-heavens.[13]

This opinion has had substantial influence and has been adopted by a number of books, because it is from an authoritative masterpiece in the history of Chinese astronomy. Today, it seems too aggressive due to its time, and thus unfair. Still, it offered a speculation about the time of the

---

[12] Yamada Kyoko. "Theory of Canopy Heavens of Emperor Wu of the Liang Dynasty and the Universe Garden", in his collection "Ancient Philosophy and Science and Technology Culture in East China", Liaoning Education Press, (1996), p. 165.

[13] *The History of Chinese Astronomy* compiled by the research team of Chinese astronomy history, p. 164.

Changchun Hall Conference — 525 A.D., or the sixth year of the Putong era, though also without any supporting evidence like Yamada Keiji's guess.

All the opinions above have their own value, but they all fail to probe into the background and significance of the Conference. Here, there are two important questions we have to figure out:

1. What is the content of the cosmic theory Emperor Wu argued for in the Conference and the theory's relationship with Indian astronomy?
2. Why *Astronomical Treatise, the Book of Sui* said that Emperor Wu "expressed views exactly the same as *Zhou Bi Suan Jing*" in the Conference?

The first question is easy to answer, for the main content of the emperor's speech is recorded in the first volume of *Kai Yuan Zhan Jing*.[14] In the very beginning, Emperor Wu dismissed completely all other cosmic theories with this exaggerating rhetoric:

> *Since ancient times, there has been various theories about heaven, all made-up nonsense by people who couldn't read astronomical phenomena. When each school has its own theory and everyone holds their own opinion, the disparities are huge and the errors shocking. How can you expect to see the sky with a basin over your head? It's like trying to talk about the vastness of the sky while living on the tiny antenna of a snail, or hoping to fathom the sea with the shell of a clam. That's pure madness!*

Such an assertion is exceedingly bold and arbitrary indeed. What is especially notable is that then the theory of sphere-heavens had already gained the upper hand and been accepted by most astronomers. However, Emperor Wu hastily negated it without giving any astronomical evidence, which should be hardly commendable but for his authority as the supreme ruler. Also, the model of the universe he himself advocated for was also offered as a last resort without being borne out by solid evidence:

---

[14] Also Tian Xiang Lun (On Celestial Phenomena) in Quan Liang Wen (Vol. 6), with words slightly different from the *Kai Yuan Zhan Jing*.

There is a mountain called the Jingang Mountain, or the Tiewei Mountain, outside the four seas. To the south of the Jingang Mountain lies the Black Mountain, around which the sun and the moon circle, completing a cycle every day and night. (When the sun and the moon) travel to the south of the Black Mountain, they become visible; when they travel to the opposite side, they become invisible. In winter, the sun travels in a low position and in summer, high; correspondingly, days are short when the sun travels low, and long when it travels high. Hence, the four seasons and the day-night cycle come into being.

Such a model of the universe and its explanation of the seasons are preposterous for many sphere-heavens advocates. But Emperor Wu was not making all it up without foundation — from the third section of this chapter, we already knew that he had drawn on the ancient Indian cosmology in the Buddhist texts.

The second question has to be answered based on the first question. If Emperor Wu argued for an Indian universe model, why did *Astronomical Treatise, the Book of Sui* say that he "expressed views exactly the same as *Zhou Bi Suan Jing*" in the Conference? It is unconvincing to say that the emperor was trying to revive the theory of canopy-heavens, because both the canopy-heavens and the sphere-heavens were dismissed by him. Again as we already knew with reference to the third section of the chapter, the model of the universe described in *Zhou Bi Suan Jing* was exactly originated in India! So that record in *Astronomical Treatise, the Book of Sui* is actually a completely veracious account — just omitting the reasoning.

8. Emperor Wu of Liang and the Tongtai Temple

Discussing the relationship between the Tongtai Temple and Emperor Wu, most scholars only focused on the emperor's forsaking of secularity for a monastic life there. But in recent years, the Japanese scholar Yamada Keiji wrote a long article and in it made the most valid argument that the architecture of the Tongtai Temple was modeled on the Buddhist universe.[15]

---

[15]Yamada Keiji (Japanese). "Emperor Wu of Liang's Theory of Canopy-heavens and World Garden", *Ancient East Asian Philosophy and Technology*, p. 165.

The most detailed record about the Tongtai Temple can be found in Volume 17 of *Jian Kang Shi Lu (True Records of Jiankang)*:

In the first year of the Datong era, ... Emperor Wu of Liang built the Tongtai Temple behind the imperial palace. From the palace, a new gate called the Datong Gate led to the south entrance of the temple. The temple was named "Tongtai" to suggest "harmony and peace". In the day or at night, Emperor Wu would frequently pass the gate to study Buddhist doctrine. The temple was located about 3 km east of the county. (According to Yudizhi, the temple lay west of the Beiye Gate and spread to the west of the main road inside the Guangmo Gate. It was built by order of Emperor Wu during the Putong era, which is from 520–526 A.D. The place was the back garden of the imperial palace in the times of the Wu state, and where the main judicial officials worked in the Jin Dynasty. Later, the judicial officials moved outside the palace, and the place was turned into a temple. The temple, surrounded by a moat, had a nine-story pagoda, six major halls, and over ten minor halls, its rooms all imitating the shape of the sun and the moon. There were also places for Buddhist meditation in a little wood. A three-story Bore Platform lay in both the eastern and Western side of the wood. The northwest part was occupied by the Bodian Hall. And in the southeast lay the Xuanji Hall; outside the hall, rocks were accumulated while trees were being planted to form a hill and a gaitianyi [an armillary instrument illustrating the canopy-heavens theory], driven by water, was in motion. About 10 years after the temple was built, a great fire broke out, engulfing almost everything and leaving only the armillary instrument and the Bodian Hall. Rebuilding work was launched immediately and a new twelve-story pagoda was added to the plan. But before its completion, Hou Jing declared a rebellion and imprisoned Emperor Wu who was then starved to death.) The first time Emperor Wu visited the Tongtai Temple, he offered himself to the service of the Buddha (a gesture meaning he was renouncing the throne in favor of a monastic life in the temple) — which, of course, was followed by the ministers' begging and the king's finally agreed return — and made the eighth year of the Putong era the beginning of the Datong era.

The above record is of great importance. First, it identified the general period when the temple was built. About this, Volume 1 of *Xu Gao*

Seng Zhuan (Further Biographies of Eminent Monks) can also offer reference:

> *In the first year of the Datong era, the Datong Gate was constructed in the north of the palace and the Tongtai Temple was built. The constructions in the temple followed the style of the palace. The towering nine-story pagoda reached to the sky and the beautiful garden was like a wonderland.*

Accordingly, we can arrive at the conclusion that the Tongtai Temple was constructed in the first year of the Datong era.

By the way, Emperor Wu had offered himself multiple times to the service of the Buddha in the Tongtai Temple, and this is often seen as evidence of his extreme Buddhist obsession. But to do him justice, Emperor Wu was not the only one at that time to make such offerings as the leader of an empire. For example, Emperor Wu of Chen and the Final Lord of Chen had also manifested the same behavior. Volume 19 of *Jian Kang Shi Lu* said:

> *In 558 A.D., ... one day in the fifth month (by lunar calendar), Emperor Wu of Chen visited the Dazhuangyan Temple and offered himself to the service of the Buddha. The next day, the ministers all petitioned their Majesty to return to the palace.*

And Volume 20 "Hou Zhu Chang Cheng Gong Shu Bao" of the same book said:

> *In the ninth month of 582 A.D. (the 14th year of Taijian Period) the emperor presided a grand ceremony in front of the Taiji Hall and announced he had become a monk. He presented his carriages and garments to the Temple and proclaimed a general amnesty.*

These acts of offering oneself to the service of the Buddha seem more like symbolic gestures, and thus are certainly not true renunciation of imperial privileges.

The relationship is obvious between the architectural style of the Tongtai Temple and the ancient Indian universe model that Emperor Wu advocated for in the Changchun Hall Conference. Yet there is something more to delve into in the above cited record from *Jiankang Shilu*.

"In the southeast lay the Xuanji Hall; ... a *gaitianyi* [an armillary instrument illustrating the canopy-heavens theory], driven by water, was in motion". This record, which should be describing an Indian Buddhist universe model, is extremely important; "gaitianyi" has not been seen ever in the traditional Chinese *tianxue* instruments. Actually, the whole Tongtai Temple is a giant, emblematic "gaitianyi". As to the *gaitianyi* in front of the Xuanji Hall, it merits further exploration from the perspective of Chinese *tianxue* instrument history.

9. Why did Emperor Wu of Liang reform the water clock system

The Liang Dynasty did not develop a new calendar, as Volume 147 of *Zizhi Tongjian* said:

> [In 504 A.D.] the emperor commanded the formulation of a new calendar. Court adviser Zu Geng submitted a memorial to the throne arguing that the old calendar should not be changed according to his father, Zu Chongzhi. Five years later, the emperor ordered court historian to examine the old and new calendars; as a result, the new calendar was more accurate while the old one was comparatively crude.

In 510 A.D., Zu Chongzhi's Daming calendar was adopted. Since the old calendar could not be changed, Emperor Wu of Liang ordered Zu Geng to write *Loujing* and upgrade the *kelou* (water clock). *Astronomical Treatise, the Book of Sui* says:

> In the sixth year of the Tianjian era, Emperor Wu tried to divide the 100 *ke* according to the 12 *shichen* [dual hour, another time keeping system of ancient China], but with 8 *ke* for each *shichen*, there was still a remainder of 4. So, he changed the time system to make a day consist of 96 *ke*, and thus each *shichen* had 8 *ke*.

As we all know, *kelou* is a very important time measuring device in ancient times. And it is also an astronomical instrument which, like armillary spheres, was considered sacred. As Zu Chongzhi and his son Zu Geng were both astronomers, naturally it was up to them to craft the new *kelou*. Records about the *kelou* through the ages were mainly collected in *Astronomical Treatise*.

Upon the completion of the new *kelou*, Lu Chui, a retainer under the crown prince, composed a poem which was recorded in Volume 56 of *Wenxuan*. The prologue of the poem said:

> On the16th day of the tenth month by lunar calendar of the sixth year of
> the Tianjian era, the new kelou was finished and presented to the
> emperor. It was extremely accurate and could at the same time be
> applied to the correction of astronomical calculations, the prediction of
> weather, the monitoring of seasons and the examination of the calendar.

It praised the craftsmanship and ingenuity of the new *kelou* — though many expressions it used are eulogistic clichés. For the new *kelou*, Emperor Wu of Liang commanded that an entire day be divided into 96 *ke* (quarter-hours).

Using the *kelou* to keep track of time, ancient China would traditionally divide an entire day into 100 *ke*. As an independent time system, the 100-*ke* system should not have led to any problem. However, when another time system — the double-hour system — entered the equation, inconveniences arose as there was no simple formula enabling the conversion between the two systems. Such inconveniences had at least existed for hundreds of years before the times of Emperor Wu of Liang — only Emperor Ai of Han ever switched to a 120-*ke* system — and continued to exist after the Liang Dynasty for more than a millennium, until the spread of Western astronomy into China in the late Ming and early Qing. So here comes the question: Why would Emperor Wu of Liang want to change the 100-*ke* system into the 96-*ke* system?

Astronomical Treatise, the Book of Sui seems to give the answer: to fit in with the 12 double-hours. This could be one reason why Emperor Wu reformed the time system, but why only he thought of it, while other dynasties remained indifferent to the mismatch?

Actually, Emperor Wu's reform of the *kelou* system was related to the Indian calendar introduced to China together with Buddhism. In the foregoing discussion, we have already mentioned that because of his cultist-like worship of Buddhism, Emperor Wu wanted to replace the then-dominant "theory of sphere-heavens" with an ancient Indian model of the universe which was spread to China with Buddhism. This is an important piece of evidence that Emperor Wu cared and intervened in astronomical matters; and his reform of the *kelou* system further proves that he made interventions based on foreign astronomy.

According to Volume 11 of *Da Fang Guang Fo Hua Yan Jing* (Taisho Tripitaka No. 293):

A wise man should know that the day is divided into eight hours: four for the day and four for the night. Each of the eight hours is further quartered, leading to 32 quarters of the day measured by the water clock. Of the four hours of the day, the first starts at sunrise until midmorning; the second is from midmorning to noon, the third from noon to midafternoon and the last from midafternoon to sunset.

The above record in the sutra elaborated on a folk system for time-keeping in which the day is divided into eight hours and each hour is further quartered, making 32 quarters of an entire day. It also told us that ancient India also used water clocks to measure time. In some sutras, it is also said that the day is divided into six hours. For example, *Fo Shuo Da Cheng Wu Liang Shou Zhuang Yan Jing* (Taisho Tripitaka No. 363) said:

After I attained enlightenment, I live in a pagoda. All the Buddhas are always happy in the entire six hours of a day.

From this, we know that there were two time systems used by the ordinary people of ancient India — the eight-hour system and the six-hour system.

What's more, *Da Fang Guang Fo Hua Yan Jing* also has a record saying that:

The king is diligent and skips sleep. During the four hours of the night, he enters a state of calmness in the first two hours, rises and clears his

mind for the absorption of Buddhist wisdom in the third hour, and con-
templates over the external world and tries to liberate himself from
greed and hate in the fourth hour. When the day breaks, the king will first
chew willow branches and then continue his routine which include nine
other things: bathing, dressing, dabbing perfume, wearing headdresses,
oiling the feet, putting on the leather slippers, bringing the chatra [an
auspicious symbol in Hinduism], forming up his men with solemnity and
offering sacrifices respectively.

It can be seen that arrangements were made according to the eight-
hour-a-day time system. In the first two hours of the night, the king enters
"a state of calmness" — here, such a state should not mean sleep, because
as it is said, "the king is diligent and skips sleep". In the third hour, the
king "rises and clears his mind for the absorption of Buddhist wisdom";
this is likely to be some kind of meditation with music. In the fourth hour,
it is mentioned that the king "contemplates over the external world" —
this probably means the king is mulling over state affairs. From chewing
willow branches at daybreak to offering sacrifices, there are in total ten
things that need to be done. Each of them, according to the Buddhist
explanation, has in it ten virtues. For example, the ten virtues for "chewing
willow branches" are:

First remove undigested overnight food; second expel phlegm; third help
detoxification; fourth clear up teeth; fifth freshen breath; sixth improve
eyesight; seventh soothe the throat; eighth moisturize lips; ninth
enhance voice; and tenth improve appetite.

Some likened the act of "chewing willow branches" to tooth brushing.
It is reasonable to some extent, but when taking into account the ten
virtues, obviously it also has medical significance and is more beneficial
than tooth brushing.

The king needed to finish the ten things mentioned above within two
quarters of the first hour of the day. When the sun rises:

First [the king] sends for the physician to check his health so that he can
take some medicine accordingly in the day and at night. Then, he

summons the astronomer to see whether there's any anomaly about nature and the astronomical objects — if there is, then he has to hold a great event to ward off the misfortune. All the findings of the diviner need to be reported. If state secret is involved, then the thing can be reported to the throne through confidential ways.

After meetings with the physician and the astronomer, the king would hold court, which lasted until the latter half of the first hour after sunrise. Then in the second hour, the king would have a meal, followed by an hour for bath and recreation; in the fourth hour, he came to the grand hall to hold a meeting in which he would invite the great Samanas and Brahmans to give speeches as well as senior officials and hermetic sages to consult on national affairs.

According to the record, we can make a detailed time schedule of the Indian king:

Four hours of night	First to second hour	Enters a state of calmness
	Third hour	Clears mind for the absorption of Buddhist wisdom
	Fourth hour	Contemplates over external world and tries to be free of greed and hate
Four hours of day	First quarter of first hour	Chews willow branches
	Second quarter of first hour	Do ten things
	Third quarter of first hour	Summons the physician and then the astronomer
	Fourth quarter of fourth hour	Holds court
	Second hour	Have meal
	Third hour	Have bath and recreations
	Fourth hour	ask the great Samanas and Brahmans to give speeches
		Consult with senior officials and hermetic sages on national affairs

Obviously in the Buddhist view, a good king should follow the above time schedule strictly. Since Emperor Wu was a fervent follower of Buddhism, it is not hard to imagine that he would hold the king of ancient India, the birthplace of Buddhism, in high esteem and try to imitate his lifestyle as much as possible. According to Emperor Wu's own account in *Jingye Fu*:

> *When I haven't donned the imperial robe, ... I was used to eating all kinds of meat and knew not what vegetables taste. After I claimed the throne, everything under heaven is in my possession and delicacies of all sorts are sent to my table.... Now how can I still allow myself the luxury? So, I take vegetarian diets and refrain from meat.*

A similar record can be found in the *Annals of Emperor Wu* in *Book of Liang*:

> *[Emperor Wu] ate just one meal every day which contained no meat but bean dishes and brown rice only. With a heavy workload, he would just wash his mouth [and skip eating] if he missed his only meal at noon. He dressed like commoners, and used a mockmain bed net; he would wear a hat for three years and use a quilt for two years. On the whole, he led a thrifty life. And after 50, he abstained from sex.*

In Volume 17 of *Jiankang Shilu*:

> *[Emperor Wu] abstained from sex after he reached 59. His concubines owned no exquisite accessories, and had cut out alcohol and music.*

And also in the "11th Year of the Datong Era" section of Volume 159, *Zizhi Tongjian*, it is said:

> *[Emperor Wu of Liang] had followed the Buddhist teachings since the Tianjian era [502–519 A.D.], keeping a simple diet without any meat. He had just one meal every day, eating only brow rice and vegetable soup. If he was too busy to dine at the mealtime around noon, then he would just wash his mouth and skip eating.*

From the record above, we can see that Emperor Wu of Liang had stopped eating meat and adhered to the Buddhist doctrine "one meal a day and no eating after noon" since he claimed the throne. He would even just wash his mouth if he was too occupied with work to have the noon-time meal. But then again, the Indian king also had his meal in the second hour according to his time schedule detailed above, which was in line with the doctrine of "no eating after noon". Then reaching his fifties, the emperor took a step further by parting with his sex life. As a man in his position who possessed the wealth of a country, he still embraced such a lifestyle — this can only be explained by his devotion to Buddhism, which is so deep that it is nearly cruel.

Emperor Wu must have heard about Kanishka and Mahamayuri, two famous kings of ancient India, and so with great respect, he modeled himself on them with the ambition to turn his empire into a land purified by Buddhism. In 517 A.D., Emperor Wu issued a series of edicts to promote Buddhism (for more details, see Volume 148 of *Zizhi Tongjian (Comprehensive Mirror in Aid of Governance)*:

> *In March, [Emperor Wu of Liang] gave an order to the textiles officer, banning workers from using the patterns of immortal saints, birds and beasts on fabrics, because they would be cut apart during the tailoring process, which was against the Buddhist values. In April, the emperor issued an edict, commanding the replacement of animal sacrifices by pastries and dried meat slices, for the original practice ran counter to the Buddhist protection of animals. In October, a third royal edict was released, which, considering that dried meat slices were still allowed as sacrifices, ordered to replace them with flat bread and the rest with vegetables and fruits.*

The ban on the patterns of saints, birds, and beasts only slightly influenced the variety of clothing. But doing away with animal sacrifices was a matter with serious implications. It led to a huge conflict with the traditional Chinese views and ignited a hot debate across the country. Scholar-officials deemed it as "totally abandoning the tradition of offering sacrifices" to ancestors, and argued that when today's people abandoned the tradition, the future generations would follow suit; then how was it different than cutting off the family line? Nevertheless, Emperor Wu insisted stubbornly, banning further the use of dried meat, which he ordered to be replaced by flat bread in his October edict. Furthermore, Emperor Wu, granting the petition of some Sramanas, had also prohibited fishing and hunting in the historical places of Danyang and Langye (Volume 26 of *Guang Hong Ming Ji*). Though the prohibition was met with opposition, it was put into practice anyway. And according to Volume 149 of *Zizhi Tongjian*, "In the second year of the Putong Era [521 A.D.], [Emperor Wu of Liang] established the Gudu Garden in Jiankang [present-day Nanjing] to care for the poor". To all these, Hu Sanxing commented:

> *There was a social system intended to take care of those who had no kin and couldn't support themselves in ancient China, but Emperor Wu was not continuing this system; rather, he was just emulating Anathapindika, a great Buddhist disciple.*

From all the things detailed above about Emperor Wu of Liang, it can be seen that he was sparing no effort to make himself a good king from the Buddhist perspective.

Since Emperor Wu was determined to promote Buddhism and set an example with his own adherence to the Buddhist discipline, he naturally would follow strictly the time schedule of the Indian kings previously described. And as we have already known, the schedule was arranged according to a folk system for time-keeping (eight hours a day). So, we can infer that as Emperor Wu had willingly embraced the Buddhist time-tables, and so he would also adopt the attendant Indian folk system of time.

However, though ancient India and China both used water clocks to keep time, the Chinese divided a day into 100 *ke* while the Indian folks divided it into eight hours (or six hours). Without a consistent time system, it would be hard to follow the Buddhist timetable. So, in Emperor Wu's view, it was very necessary to switch from the 100-*ke* system to the 96-*ke* system that allowed easy conversions to the Indian time. Meanwhile, the 96 *ke* could also be conveniently converted to the 12 *shichen* (dual hour), and they did not vary much from the original 100 *ke*, so there would not be a disparity between the lengths of the *ke* after the reform. Hence, the 96-*ke* system is comparatively an ideal choice.

Some might ask if the ancient Indian kings really had to follow so strict timetables, to the extent that when Emperor Wu was emulating them, he needed to change the time system at the same time. Actually, it is not uncommon to see stringent time requirements for a king's acts in ancient cultures. For example, in ancient China, some activities of the emperors were also subject to the predictions of electional astrology. In ancient India, the timing of religious rituals required a very high degree of precision, to the extent that the determination of an accurate time almost became the sole purpose of the ancient Indian astronomy.[16] Take a religious ceremony of ancient India, which is described in Section 6, Chapter 15, Book Three of *Ramayana*:

People held the Agrayana ceremony to pay tribute to the ancestors and deities; holding the ceremony on time will help clean their sins.[17]

Āgrayana is an ancient ritual of India. Note that here it is emphasized that only holding the ritual "on time" will help clean up sins. Obviously, punctuality was of great importance. Emperor Wu admired the Buddhist country of India, and would naturally devote himself to the various Buddhist rituals. Yet the rituals were scheduled according to the Indian time system, so it seemed imperative to change the 100-*ke* system into the 96-*ke* system in order to hold the rituals on time for Buddhist blessing.

In 544 A.D., Emperor Wu replaced the 96-*ke* system with the 108-*ke* system, 37 years after its release. Conversion between the new 108-*ke* system and the six-hour system, the eight-hour system, and the 12-*shichen* system is also quite easy.

After the Liang Dynasty, the 108-*ke* system was reinstated and remained in use until the Western calendar came to China in the late Ming and early Qing dynasties. Ordinary people in the West used a 24-hour time-keeping system which matched with the 12-*shichen* system of China.

---

[16] D. Pingree. *History of Mathematical Astronomy in India*, (1981), p. 629.

[17] Valmiki (author), Ji Xianlin (translator): *Ramayana*, People's Literature Publishing House, (1992), p. 97–98.

Under such a circumstance, the 96-*ke* system was again put to use and became the official time system of the Qing Dynasty. Today, an hour consists of four quarter-hours, and a day has precisely 96 quarter-hours — this is exactly the system Emperor Wu had promoted! The emperor would never have anticipated that his reform would be reprised after a thousand years.

Notably, the two shifts from the 100-*ke* system to the 96-*ke* system in the Qing and Liang Dynasties were all influenced by foreign astronomy — respectively, European astronomy and ancient Indian astronomy introduced into China with Buddhism.

Chapter 9

Sino-foreign Exchanges in Ancient Tianxue (2)

1. Five periods of ancient Indian astronomy

We can generally be assured of the same origin for ancient Chinese and ancient Indian astronomy, considering the great similarity between the universe model in *Zhou Bi Suan Jing* and that from ancient India, though the means of cross-cultural communication has been kept in the dark. However, in some historical events, the means are specified with solid evidence. Before we delve into such incidents, let us have a brief understanding of the development of ancient Indian astronomy. After all, only against specific times and backgrounds can the incident and its significance be better taken in.

According to the prestigious scholar D. Pingree,[1] the development of ancient Indian astronomy can be divided into five phases:

I. Veda Period (c. 1000–400 B.C.). With the most dynamic environment for indigenous astronomy, the period features fresh understanding of yugas and establishment of naksatyas, which, bearing some resemblance to the twenty eight lunar mansions in ancient China,[2] represents

---

[1] D. Pingree, *History of Mathematical Astronomy in India*, (1981), p. 534.
[2] Views vary greatly as regards the origin of 28 lunar mansions. Please refer to pp. 302–313 of *Tianxue Zhenyuan* for more clues.

the location of the moon every night. Such content is mainly recorded in various Veda works.

II. Babylonian Period (c. 400–200 B.C.). During this time, lots of astronomical knowledge was introduced from Babylon to India. Babylonian astronomical parameters, mathematical models (as exemplified by the linear zigzag function), units of time, and astronomical instruments made their presence felt in Sanskrit masterpieces.

III. Greek-Babylonian Period (c. 200–400A.D.). The astronomy of the Seleucid Empire in Babylon, after the adaptation by Greeks, was introduced to India in this stage, featuring descriptions of planetary motion, geometric calculations relating to eclipses and shadow lengths, etc.

IV. Greek Period (c. 400–1600 A.D.). The real beginning of the introduction of astronomy from Greece to India was marked by non-Ptolemaic astronomy's entrance into India, which was developed from Aristotle's conceptions. Under the influence of Greek astronomy, there arose a host of distinguished Indian astronomers and astronomical classics, which fell into five major schools, namely, Brāhmapaks, Āryapaks, Ārdharatikapaks, Saurapaks, and Gansapaksa.

V. Islamic Period (c. 1600–1800 A.D.). It was a period when Indian astronomy was swayed by that of Islam, which originated from ancient Greece. Though during the last four phases, Indian astronomy received profound impacts from other cultures, its indigenous features did not fade away. As a result, when Indian astronomy was circulated to China, it was endowed with Indian characteristics on the one hand and Babylonian-Greek heritage on the other.

2. *Qi Yao Rang Zai Jue*

Qi Yao Rang Zai Jue, composed in the 9th century by a Brahman monk migrating from India to China, is an astrology guide written in Chinese and one of the world's oldest ephemerides of planets. As an outstanding classic, it has played an active role in ancient cultural exchanges between the West and East, which is intriguing to be considered in retrospect. More than four decades ago, Joseph Needham had called for a monographic study on *Qi Yao Rang Zai Jue*, being incapable to embark on such a demanding task himself due to professional limitations. Later, a Japanese scholar Michio Yano conducted research on it. Years ago, Dr. Niu Weixing, a student of mine, further unveiled this academic treasure through his exhaustive research.[3]

We can only extract scarce information from the inscription at the beginning of the work: the author is Chin Chu-cha, an Indian Brahman monk who lived in the Tang Dynasty, presumably in the first half of the 9th century. Lots of Indian or Western monks in that era are included in *Biographies of Eminent Monks*, but no signs of Chin Chu-cha can be found, and his profile remains a mystery.

Qi Yao Rang Zai Jue failed to be passed down in China. What we see now are copies made by Japanese Buddhist monks who introduced the masterpiece from China to Japan in the Tang Dynasty. Zong Rui, being one of them, brought a wealth of Buddhist tantras to Japan four years after he arrived in China in 862 A.D. These so-called "oriental tantras" have been preserved until today. *Qing Lai Lu* (*List of Borrowed Books*) records all the Buddhist classics that Zong Rui brought to Japan, among which is *Qi Yao Rang Zai Jue*.

Of all the commonly seen Buddhist canons these days, *Qi Yao Rang Zai Jue* can only be found in *Taishō Tripiṭaka* and *Pinjiazang*, which may derive from *Hongjiaozang* likewise (though the latter was completed in Shanghai in the early Republican era, it primarily consulted the Japanese KUGEU ZO). *Taishō Tripiṭaka* was published later, after correcting part of the mistakes in the scriptures. The *Qi Yao Rang Zai Jue* therein ends with a date — "Mar. 5th, the first year of Changbao Period", which is regarded as the time when the scripture was transcribed. Changbao is the title of a Japanese emperor, and the year is actually 999 A.D.

There is more than one version of *Qi Yao Rang Zai Jue* circulating in Japan. The epilogue in *Taishō Tripiṭaka*, written by a Japanese monk Kuai Dao in Hasedera Temple, reads:

Zong Rui's Qing Lai Lu listed Qi Yao Rang Zai Jue. I combine its two volumes into a comprehensive work. While collating the copy from

---

[3] Niu Weixing. *Mathematical Astronomy and Its Origin Revealed in Chinese-version Sutras — Based on Ephemerides in Qi Yao Rang Zai Jue*. Master's Dissertation of Shanghai Astronomical Observatory, CAS, (1993).

Zhucha Temple, I spotted lots of mistakes and abstruse content. So I requested the one preserved in Ninnaji Temple for proofreading, but still there remain several doubts. For a start, I mark the differences between the two texts at the beginning of the work, and have them inscribed by craftsmen. It is my wish that a more thorough version will be discovered, which people can use to avert disasters and enjoy a peaceful life.

Midsummer of Year Renxu

Year Renxu is 1802 A.D. The two-volume *Qi Yao Rang Zai Jue*, popular now, is actually amended by Kuai Dao. Though a lack of the third volume, the work is complete in content as an astrology guide, even if there was Vol. 3, it could be an appendix, with little effect on completeness.

3. Prayers in Buddhist tantras become scientific heritage

Buddhist tantras lay great emphasis on *tianxue* and devote much of the content to a magical practice that averts disasters or solicits blessings in compliance with the motion of celestial bodies. *Qi Yao Rang Zai Jue*, as its name indicates, is a guide on such a practice through observation of Qiyao (seven luminaries), namely, the sun, the moon, and the five planets.

Vol.1 begins with "Zhanzai Rangzhi Fa" (the method of divining and averting misfortunes). In the order of the sun, the moon, Jupiter, Mars, Saturn, Venus, and finally Mercury, it lists different kinds of good or bad luck, respectively, brought by the seven celestial bodies when they move to people's Minggong (the constellation determined at birth). The following content pertains to Jupiter:

Jupiter is the son of the Oriental God Qingdi, completing one cycle around the sun every 12 years. When it moves to one's Minggong in the spring, the person will attain both promotion in the officialdom and wealth; when it moves to one's Minggong in the summer, the person will easily produce an offspring; when it moves to one's Minggong in the autumn, the person is inclined to having diseases or injuries; when it moves to one's Minggong in the winter, the person will be blessed with a large fortune; when it moves to one's Minggong during the period of

"Ji"(18 specific days in a season), the person will have false news and quarrels.

Should disasters arrive when Jupiter moves to one's Minggong, the person ought to make a drawing of a human-like god with a dragon head and an ethereal outfit which changes its color in different seasons, and then wear it on the neck. When Jupiter shifts to Minggong mansion, discard the drawing in the well and everything will turn better.

The succeeding three parts all revolve around luminaries. In Part II, "Jiuyao" (nine luminaries) includes another two imaginary planets — Rahu and Ketu besides the aforementioned seven luminaries. The two are very important in *Qi Yao Rang Zai Jue*, which I will probe into in the following sections. In about 1000 A.D., Buddhist tantras had gained greater influence in Japan after they were introduced from China more than 100 years ago. This can be well testified by wide use of the *Qi Yao Rang Zai Jue*, now contained in *Taishō Tripitaka*. Ephemerides in the scripture, as a periodic tool that can be repeatedly used, have been marked with as many Japanese regnal years as 27, the earliest one being in 973 A.D. while the latest one is 1132 A.D., as well as continuous Gan-zhi years (years designated by heavenly stems and earthly branches) to 1170 A.D.

The fundamental part in *Qi Yao Rang Zai Jue* is a series of ephemerides, which list at length the apparent movements of celestial bodies in a cycle, that is, a specific time period chosen according to their motional law. So theoretically speaking, ephemerides can be used for a long time since celestial bodies repeat the same motion after a cycle ends. In the case of planets, synodic cycle is usually first considered, which is subdivided into minor stages including progradation, standstill, retrogradation, and disappearance. Here is an example to describe Jupiter's synodic cycle:

Jupiter moves this way during a synodic cycle: at first, it arises in the east morning sky, shifting one degree in its prograde orbit after six days and 19 degrees after 114 days; after that, it comes to a standstill for 27 days, and retrogrades for one degree in 7.5 days and 11 degrees in 82 days. The retrogradation is followed by another round of standstill; then it goes on prograde motion, shifting 19 degrees after 114 days. After occurring at dusk, it remains unseen in the west sky for 32 days, following which a new cycle begins.

These accounts and expressions are akin to those in traditional Chinese calendars. Synodic cycle is a rather short period of time employed by ancients to describe the motion of planets. However, these short periods of time can be combined to form an extended cycle — since the synodic cycle may not be precisely one year but describing celestial motion needs specific dates, months, and years, the combination of short periods is a requisite for producing a cycle of full years. *Qi Yao Rang Zai Jue* ascertains the extended cycles of five planets as follows:

Jupiter: 83 years (794–877 A.D.)
Mars: 79 years (794–873A.D.)
Saturn: 59 years (794–853 A.D.)
Venus: 8 years (794–802 A.D.)
Mercury: 33 years (794–827 A.D.)

The first extended cycle starts from 794 A.D., a time chosen as an epoch. *Qi Yao Rang Zai Jue* offers detailed records of planetary motion during these cycles, by virtue of which modern researchers are able to reproduce the then astronomical phenomena in line with the law of celestial mechanics and, furthermore, verify the accuracy of ephemerides in this masterpiece. Research shows that in the first cycle, the positions of planets as revealed in ephemerides largely conform to those in real astronomical phenomena. In the ensuing ones, however, errors gradually accumulate and the accuracy declines, something impossible to avoid in ancient times. But in view of so many marks of Japanese years in *Qi Yao Rang Zai Jue*, Japanese astrologists seem to neglect these errors — actually, for the purpose of averting disasters, the deviations from actual astronomical phenomena can be put aside.

The unique value of these ephemerides on science history lies in the fact that they have been, to date, one of the only two kinds of ephemerides from ancient China that calculate planetary positions year by year. Traditionally, Chinese astrological classics only give the dynamic table of planetary positions during one synodic cycle, a case in point being Bu Wu Xing in *Records of Calendars* of official history books, which requires further calculations if the position of a planet is demanded at a certain time. The Mawangdui silk manuscript *Wu Xing Zhan* contains such tables, but the cycles therein are rather short and incomplete. The most complete and elaborate data in *Wu Xing Zhan* pertain to Venus, including its cycle of eight years, which coincidentally is in agreement with that given by *Qi Yao Rang Zai Jue*.

Apart from ephemerides of planets, ephemerides of Rahu and Ketu are also essential in *Qi Yao Rang Zai Jue*. These are two imaginary celestial bodies in ancient Indian astronomy, the so-called "Yinyao". Their estimated cycles in *Qi Yao Rang Zai Jue* are separately 93 years and 62 years, all starting from 806 A.D. What deserves a special mention here is that ephemerides of Rahu and Ketu help to change a long-time misunderstanding in China. Previous definitive domestic monographs unanimously deem the two as ascending and descending nodes on moon's path (lunar orbit projected on the celestial sphere), namely, two intersections of the moon's path and ecliptic, one formed when the moon's path crosses the ecliptic from south to north and the other while crossing from north to south. Reasons that lead to such a misunderstanding can be traced back to ancient Chinese classics. Though it diverges far from the real meaning of Rahu and Ketu in ancient Indian astronomy, the misunderstanding is deeply ingrained[4] owing to wide popularity and no objections. Nevertheless, according to related ephemerides and explanations in *Qi Yao Rang Zai Jue*, there is no doubt that Rahu is the ascending node on moon's path while Ketu is the apogee (the point where the moon is the furthest away from the sun) on the moon's path.[5]

4. From Babylon to India to China to Japan

Though *Qi Yao Rang Zai Jue* is attributed to a Brahman monk from India, it is not a work based solely upon ancient Indian astronomy.

---

[4] For instance, the mistake is found in p. 135 of *The History of Chinese Astronomy* compiled by the research team of Chinese astronomy history, p. 513 of *Astronomy, Encyclopedia of China* published by Encyclopedia of China Publishing House in 1980 and pp. 12 and 137 of Vol. 4, *Science and Civilization in China* by Joseph Needham.

[5] For detailed explanation, please refer to Niu Weixing. *A Probe into Rahu and Ketu's Astronomical Meaning* [J]. *Astronomical Journal*, Vol. 35, Issue 3, (1994).

In the planetary ephemerides are listed the cycles of outer planets, which equal the sum of their synodic cycle and sidereal cycle. Taking Jupiter as an example, the cycle (83 years) is derived this way after adding up its synodic cycle (76 years) and sidereal cycle (7 years). All these data have some connection with the masterpieces of Brahmagupta, an ancient Indian astronomer of Brahman School. From another perspective, however, such combined cycles fall into the expertise of Babylonian astronomers living in the Seleucid Empire (312 B.C.–64 A.D.). As a matter of fact, many planetary-motion-related data in India astronomy originate from Babylon.

In describing the motion of planets during a synodic cycle, *Qi Yao Rang Zai Jue* starts with the first appearance of planets in the east sky, as in conformity with the practice in ancient Babylon, Greece, and India. Part of the data in the masterpiece, even some particular figures, can be found in astronomical literature introduced from ancient Babylon and Greece to India. A host of clues point to a fact that in *Qi Yao Rang Zai Jue*, there is a line of inheritance:

Babylon→India→China→Japan

The line has transcended time and space to become the most spectacular feature in the scientific and cultural exchanges between the ancient Oriental and Occidental world.

Qi Yao Rang Zai Jue is not an isolated example. In previous studies, we have researched many cases that indicate such an inheritance. Here, I will give a few of them: in several distinguished Sui-Tang-Dynasty calendars, we have discovered traces of Babylonian mathematical astronomy in the Seleucid Age,[6] including quintessential mathematical methods like zigzag function and second-order difference, as well as such contents as the theory of solar motion, the theory of planetary motion, celestial coordinates, lunar

---

[6] For detailed explanation, please refer to the following theses by Jiang Xiaoyuan, including *The Connection Between Babylonian and Chinese Astronomy in Light of Solar Motion Theory* [J]. *Astronomical Journal*, Vol. 29, Issue 3, (1988), *Babylon & Planetary Motion Theory in Ancient China* [J]. *Astronomical Journal*, Vol. 31, Issue 4, (1990) and *Several Questions on the History of Babylonian and Chinese Astronomy* [J]. *Journal of Dialectics of Nature*, Vol. 12, Issue 4, (1990).

motion, Zhirun cycle (the cycle of leap year) and calculations of day length; some innovative reforms in Yuanjia Calendar (443 A.D.) can find their counterparts in Indian astronomy,[7] since its author He Chengtian (370–447 A.D.), an astronomer in the Southern Dynasties, once studied Indian calendars under the guidance of Xu Guang and Shi Hui; the so-called Tianzhu Sanjia in the Tang Dynasty were, in truth, Indian astronomers migrating to China, who either assumed important hereditary positions in imperial tianxue academy or made their findings a paradigm followed by official tianxue academies and calendars.[8] All such examples show us the social context then — cross-cultural exchanges on astronomy.

5. Activities of Yelü Chucai and Qiu Chuji in Middle Asia

The rise of Mongol Empire spanning Eurasia brought about a merging of multiple nationalities and cultures, as well as a new climax in astronomical exchanges between China and other countries. Though Chinese and foreign scholars have touched upon the communication between Chinese and Islamic astronomy during that period, we still lack clear clues and conclusions on quite a few issues. In this section and also ensuing ones, I will explore six important topics in chronological order.

First, let us look into the activities of Yelü Chucai and Qiu Chuji in Middle Asia, which bear much significance despite being neglected by our predecessors.

Yelü Chucai (1189–1243 A.D.), a native of Qidan [Khitan], an ancient nationality in China, was a lineal descendant of the royal family of the Liao Dynasty (907–1125 A.D.). At first, he served in the Jin Dynasty (1115–1234 A.D.) as an official, but was later summoned to serve for the Mongol Empire. In 1219, he was appointed astrological-medical consultant of Genghis Khan and followed the soldiers in their expedition to Western regions. On the journey, he had a dispute with

---

[7] Please refer to Niu Weixing, Jiang Xiaoyuan. *The Connection Between Indian Astronomy and He Chengtian's Reforms on Calendar* [J]. *Journal of Dialectics of Nature*, Vol. 19, Issue 1, (1997).

[8] Please refer to *Tianxue Zhenyuan*, 2007, pp. 361–370.

Islamic astronomers about lunar eclipse, which is recorded in *Biography of Yelü Chucai, History of Yuan*:

> *Calendar experts in Western regions reported to the court: In May when the moon is full, there will be a lunar eclipse. Chucai didn't agree, and it turned out that no eclipse happened at all. In the next October, Chucai predicted a lunar eclipse but the Western experts voiced their doubts. When the time came, a lunar eclipse did occur with 80% of the moon disappearing.*

This incident happened in 1220 A.D., the second year after Genghis Khan led the expedition to Western regions, as can be inferred from one of the records in Vol.1 of *Records of Calendar, History of Yuan* —"In Year Gengchen, Emperor Taizu[9] led expeditionary forces to Western regions. During the journey, a lunar eclipse was predicted to occur in May but the assumption turned out to be false...." From *Journey to Western Regions,*[10] a book penned by Yelü Chucai to record this experience, we can also know that it happened in Samarkand, a place in the present-day Uzbekistan.

Yelü Chucai was versed in traditional Chinese astronomy. In the early years of the Yuan Dynasty (1271–1368 A.D.), the Jin-Dynasty Daming Calendar was still in use. However, errors repeatedly arose, one of them being the aforementioned failure in predicting lunar eclipse. Noticing that, Yelü Chucai compiled Year-Gengwu Yuan Calendar during the westward expedition (as recorded in Vols. 5 and 6 of *Records of Calendar, History of Yuan*), which, for the first time, dealt with the time gap caused by different geographical longitudes. This can be considered as one instance of the influence of Western astronomical methods on traditional Chinese astronomy system — the problem of the time gap had already been solved in ancient Greek astronomy, and in Islamic astronomy as well, which shared the same origin.

---

[9] In the original text is Emperor Taizong instead of Emperor Taizu. But during the reign of Emperor Taizong there is no Year Gengchen. So correction is made here according to p. 3330 of *Collection of All-Dynasties Records of Astronomy and Calendars*. Zhonghua Book Company, Vol. 9, (1976).

[10] *Journey to Western Regions.* Xiang Da (collator). Zhonghua Book Company, (1981).

According to other historical records, Yelü Chucai was also a master of Islamic calendars. Vol. 9 "Madaba Calendar" of *Nan Cun Chuo Geng Lu (Discontinuing My Farming)*, a book by Tao Zongyi in the Yuan Dynasty states:

> *Yelü Chucai is adept at astronomy, calendar, astrology, divination, mathematics, music and Confucianism. He has a good grasp of all foreign books. It is often said that records of planetary motion in Western calendars are more accurate than in Chinese calendars. So Yelü Chucai creates Madaba Calendar modeled after Islamic calendars.*

Considering that Yelü Chucai gained the upper hand in two arguments with calendar experts from Western Regions, we can assume that he knew about both traditional Chinese astronomy and Islamic astronomy. In that way, he could prevail over his opponents.

At about the same time that Yelü Chucai followed Genghis Khan to Western regions, another famous figure Qiu Chuji (1148–1227 A.D.) was on his way to Middle Asia, and he took the imperial order to preach Taoism for Genghis Khan. Qiu Chuji reached Samarkand in late 1221, just about the time when Yelü Chucai arrived. There, Qiu discussed with local astronomers the partial solar eclipse that had happened on May 23rd that year. Vol. 1 of *Changchun Zhenren Xi You Ji (Journey to West of Changchun Taoist)* records the anecdote:

> *Qiu arrived in Xiemisigan (present-day Samarkand)...When a local astronomer was around, Master (Qiu Chuji) asked about the solar eclipse in May. The man replied, "The eclipse took place in Chen Period (7 a.m. to 9 a.m.) and 60% of the sun is shrouded". Master said, "Earlier when we were at Luju River, we saw total eclipse at the noon. Later we went southwest to Jinshan, but local people said the eclipse had taken place in Si Period (9 a.m. to 11 a.m.), with 70% of the sun in sheer darkness. As I see it, total eclipse happened only in one place, and people saw different scenes as they were in other regions, some of which are even thousands of miles away. It is just like a fan blocking the light from a lantern, which cast a shadow in certain places where no light is available. But when you get farther away, there will be more beams visible.*

Qiu was 73 years old then, but he still managed to probe into astronomical issues during the tiring journey, from which we can get a hint of his strong interest in astronomy. His explanation and metaphor about different eclipse scenes at different geographical locations are insightful.

It does not seem like a coincidence that Yelü Chucai and Qiu Chuji both contacted and exchanged views with astronomers in Samarkand. One and a half centuries later, the place became the capital of a new dynasty — the Timurid Dynasty (1370–1507 A.D.), where a magnificent observatory had been built by the time Ulugh Beg ascended the throne in 1420 A.D. As instructed by Ulugh Beg, observation was carried out in the observatory for compilation of what is known as Ulugh Beg Astronomical Table, which was the only star catalogue in more than 1,000 years that is created independently, after the one made by Ptolemy.[11] So we can say that Samarkand seems to have a long tradition of vigorous astronomy studies.

6. Who was the Chinese scholar at Maragha Observatory?

In mid-thirteenth century, Hulagu (or Hulegu), grandson of Genghis Khan, launched warfare to conquer the Western regions. Baghdad was occupied in 1258 A.D., followed by the overthrowing of Abbasid Caliphate and the rise of the Ilkhanate (1256–1335 A.D.). With the assistance of Nasir al-Din al-Tusi, a prestigious Islamic scholar, Hulagu shifted the focus from military accomplishments to civil administration, building an observatory in Maragha, capital of the Ilkhanate (present-day southern part of Tabriz in Northwest Iran) in 1259. The world-class observatory boasted a grand scale, sophisticated equipment, and a collection of 400,000 books. As the then academic center of the Muslim world, it attracted a host of researchers from around the world.

In his *Introduction to the History of Science*, G. Sarton, honored as Father of the Science History, claims that a Chinese scholar once conducted

---

[11] Ptolemy's star catalogue is included in *Zhi Da Lun* (*Almagest*). Instead of relying on independent observations, subsequent star catalogues in the West are merely revisions of Ptolemy's finding, with some corrections made on contents like precession. And it is believed by many that Ptolemy derived his star catalogue from the catalogue of his predecessor, Hipparchus.

research at Maragha Observatory.[12] Thereafter, the topic has been frequently brought up by Western scholars, but the scholar's identity still remains as a mystery. As a matter of fact, Sarton's claim comes from C.M.D' Ohsson's *Histoire des Mongols,*[13] which states that one Chinese astronomer — Fao-moun-dji — followed Hulagu to Persia and worked at Maragha Observatory. Since accounts of his life elude us, we can only speculate the astronomer's name according to the transliteration "Fao-moun-dji", like the one adopted by Joseph Needham, i.e., "傅孟吉".[14] A deeper probe will find that the statement of D' Ohsson is based on *The Garden of Masters*, a 1317 A.D. Persian-written chronicle in nine volumes, among which the eighth one dwells on the history of China. In the chronicle is a record on this topic:

It was not until the Hulagu era that Chinese scholars and astronomers began to come here (Iran) in company with Hulagu. Among them was a master named Tu Michi ("屠密迟"), who imparted knowledge of Chinese astronomy to Nasir al-Din al-Tusi when the Islamic scholar was ordered to compile Ilkhanid Astronomical Table. When the ruler Ghazan Mahmud Khan issued an imperial edict of compiling *Biography of the Praised Ghazan*, Rashid al-Din, the prime minister then, summoned two renowned Chinese scholars, namely, Li Dachi ("李大迟") and Ni Kesun ("倪克孙") for assistance. The two were proficient in medicine, astronomy and history, and brought with them a wide range of books on these disciplines. Additionally, they expounded on the way of numbering years in China and its instability.[15]

As regards the Chinese scholar at Maragha Observatory, the above paragraph is so far the earliest historical record. Such Chinese names as "屠密迟", "李大迟", and "倪克孙" are all transliterations from Persian, and there is no way of determining their real identities. In all likelihood, "Tu Michi" is "Fao-moun-dji" ("傅孟吉"). The record also sheds some light on Chinese scholars' significant contribution in compiling the

---

[12] G. Sarton. *Introduction to the History of Science*. W. & W., Baltimore, Vol. 2, (1931), p. 1005.

[13] D'Ohsson. *Histoire des Mongols*. Feng Chengjun Trans. Zhonghua Book Company, Vol. 2, (1962), p. 91.

[14] Joseph Needham. *Science and Civilization in China*. Science Press, (1990), p. 226.

[15] *References on the History of Ancient Chinese Dynasties*. Han Rulin. Zhonghua Book Company, Vol. 6, (1981) (the Yuan Dynasty), p. 258. Some adjustments are made on the transliterated Chinese names.

Ilkhanid Astronomical Table (originally entitled ZijIl-Khani in Persian), which is the paramount accomplishment of the Nasir al-Din al-Tusi at Maragha Observatory. This can be regarded as an instance of China's impact on the West in terms of exchanges on astronomy, a consolation to those Chinese-culture advocates mentioned before.

Owing to several rounds of transliteration, the pronunciation of such Chinese names has been tremendously distorted. To determine who "Tu Michi" and "Fao-moun-dji" are, the only way out is unearthing more Chinese-written historical materials.

7. Bilingual astronomical works

Quoting Wagner, Joseph Needham once discussed two transcripts on astronomy preserved in the Russian Pulkovo Observatory. The two have the same content, namely, a table starting from 1204 A.D. to record the motion of the sun, the moon, and five planets, and they were both completed in 1261 A.D. We should, however, note that the pair of transcripts are in different languages, one in Persian and the other Chinese.

Since 1261 A.D. was the second year to mark Kublai Khan's ascension to the throne, Dr. Joseph Needham made an assumption that the two transcripts resulted from the cooperation between Jamal ad-Din and Guo Shoujing. Pulkovo Observatory was burnt during the WWII, Joseph Needham could only hope that they wouldn't burn to ashes.[16]

Before Joseph Needham, G. Sarton had researched on another bilingual astronomical work during this period, which is a masterpiece of Ata ibn Ahmad al-Samarqandi, an Islamic astronomer. Containing a table of lunar motion, it was presented to a Yuan-Dynasty prince. The manuscript is now preserved in Paris, and in photocopies revealing its partial content given by G. Sarton, we can discover Mongolian annotations beside the main body (in Arabic) and Chinese characters on the front page.[17] The Yuan-Dynasty prince, mentioned earlier, is said to be Aratnasili, one of the lineal descendents of Genghis Khan and Kublai Khan.[18]

---

[16] Joseph Needham. *Science and Civilization in China*, Vol. 4, (1990), p. 475.

[17] G. Sarton. *Introduction to the History of Science*, Vol. 3, (1947), p. 1529.

[18] Joseph Needham. *Science and Civilization in China*, Vol. 4, (1990), p. 475.

In 1267 A.D., seven years after Kublai Khan's enthronement, Jamal ad-Din, a 13th-century Persian astronomer, offered seven Western astronomical instruments to the emperor as a tribute. The original names of the seven instruments and their translations, along with the shapes and functions are illustrated in *Records of Astronomy, History of Yuan*, which have aroused great interest among Chinese and foreign scholars. Since these instruments do not exist anymore, scholars hold slightly different opinions towards their properties and functions. Here, I would like list their transliterations, related Arabic names set by W. Hartner, and functions (according to *Records of Astronomy, History of Yuan*) while briefing on the conclusions of major researches.

8. Jamal ad-Din and his seven astronomical instruments

(1) "Dhatu al-halaq-I is the equivalent of what Chinese call as an armillary sphere". Dr. Joseph Needham thinks it is an equatorial armillary sphere, while Chinese scholars consider it as an ecliptic armillary sphere.[19] It is a traditional instrument for astronomical observation in ancient Greece.

(2) "Dhatu' sh-shu 'batai is the equivalent device in China which measures celestial stars and luminaries". Both Chinese and foreign scholars tend to regard it as the organ on parallactic on mentioned by Ptolemy in *Almagest*.[20]

(3) "Rukhamah-i-mu '-wajja is an instrument for measuring the shadow cast by the sun to predict equinoxes". It is used to determine the precise time of the Spring Equinox and the Autumn Equinox after being placed in an airtight room which has only one crevice on the ridge in the direction of due west to east.

(4) "Rukhamah-i-mustawiya is an instrument for measuring the shadow cast by the sun to predict solstices". Much similar to the third instru-

---

[19] Please refer to p. 200 of *The History of Chinese Astronomy* compiled by the research team of Chinese astronomy history.

[20] Ptolemy. *Almagest*. V, 12. & Joseph Needham. *Science and Civilization in China*. Vol. 4, (1990), p. 478.

ment, it is used to determine the precise time of the Summer Solstice and the Winter Solstice, functioning together with a room which has only one crevice on the ridge in the direction of due north to south.

(5) "Kura-i-sama is the equivalent of what Chinese call as 'Hun Tian Tu'". A consensus has been reached among Chinese and foreign scholars — it was a celestial globe found in both China and the West in ancient times.

(6) "Kura-i-ard is the equivalent of what Chinese call as 'Di Li Zhi'". It is a terrestrial globe, and it is unanimously recognized by Chinese and foreign scholars.

(7) "Al-Usturlab is a time-measuring instrument for the day and night". Actually, it is an astrolabe popular in Arabic countries and Europe in the Middle Ages.

Of the above seven instruments, (1), (2), (5), and (6) had already been adopted in ancient Greek astronomy before being brought to China. It was passed down for generations and even inherited by Arabic astronomers. (3) and (4) had quintessential Arabic features while (7), the astrolabe, originating from ancient Greece, became one of the distinctive features of medieval Arabic astronomy — the sophisticated astrolabes made by Arab craftsmen had long enjoyed a good reputation. Thus, the introduction of these seven instruments to China is laden with much significance.

Four years after the gifting, Kublai Khan demanded the establishment of Huihui Astronomical Observatory in Shangdu (present-day southeastern part of Duolun County, Inner Mongolia) and appointed Jamal ad-Din as the head. When the Yuan Dynasty was overthrown and Ming soldiers occupied Shangdu, key officials in the Observatory were dispatched to the then capital (present-day Nanjing) to serve the new dynasty. However, the whereabouts of the astronomical instruments became a mystery, which no records so far have been able to trace. Given that instruments in the official astronomical bureau of Yuan have all been delivered to Nanjing, there are scholars who suppose that those in Huihui Astronomical Observatory might have undergone the same experience. Nevertheless, from my vantage point, there is a slim possibility that things like organ on parallel action and the two rooms for predicting equinoxes and solstices could be transported.

Even now, scholars know very little about this Jamal ad-Din. Most Chinese scholars accept the proposition of Dr. Joseph Needham, regarding him as an astronomer once working at Maragha Observatory, who later was sent to serve Kublai Khan by Hulagu Khan (brother of Kublai Khan) or his descendants.[21] A recent study reports that Jamal ad-Din is, in truth, what Rashid (i.e., the aforementioned Rashid al-Din) recorded in the masterpiece *Jami'al-Tawarikh* as Jamal al-Din. The very person came to China between 1249 and 1252 A.D., first served Möngke Khan, and then Kublai Khan. After Kublai Khan ascended the throne, he sent Jamal ad-Din back to Maragha Observatory in Ilkhanate for research and study, and it was not until 1267 A.D. that Jamal ad-Din returned with his new accomplishments (seven Western astronomical instruments and Perpetual Calendar).[22]

9. Foreign astronomical books at Huihui Astronomical Observatory

Huihui Astronomical Observatory in Shangdu is of vital importance in the history of Islamic astronomy — this is unarguable in view of its close relationship with Maragha Observatory of the Ilkhanate, the leadership of Jamal al-Din, and its central focus on Islamic astronomy. It can be compared to a bridge connecting Maragha Observatory and Samarkand Observatory of the subsequent Timurid Dynasty (1370–1506 A.D.). Besides, it played a pivotal role during exchanges between Chinese and Islamic astronomy, for which the following account in Vol. 7 of *Mi Shu Jian Zhi (Historical Documents of Mishujian)* can offer a clue:

> *On the 18th of the sixth lunar month in 1273 A.D., an imperial decree came, ordering that Huihui Astronomical Observatory and Han'er (Beijing) Astronomical Observatory be both administered by Mishujian.*

In totally different astronomical systems, the two observatories, however, were led by the same administrative organ — Mishujian. This interesting

[21] Please refer to p. 199 of *The History of Chinese Astronomy* compiled by the research team of Chinese astronomy history.

[22] Li Di. *Nasir al-Din al- Tusi and China* [J]. *Materials on Science History in China*, Vol. 11, Issue 4, (1990).

phenomenon is rarely seen in the history of astronomy (if not the only one), and its possible impact will be discussed later in this book.

Unfortunately, we have acquired very limited information about this significant observatory. Of all that we know, a booklist in Vol. 7 of *Historical Documents of Mishujian* is worth our attention — all the books listed were once enshrined in Huihui Astronomical Observatory, among which 13 works are related to astronomy. The books are as follows:

(1) *Si Bo Suan Fa Jia Shu*, by Wuhulie, 15 sections.
(2) *Yun Jie Suan Fa Jia Mu*, by Hanli Suku, 3 sections.
(3) *Zhu Ban Suan Fa Ji Mu Bing Yi Shi*, by Safinat Handasiyah, 17 sections.
(4) *Zao Si Tian Yi Shi*, by Maizhesi, 15 sections.
(5) *Jue Duan Zhu Ban Zai Fu*, by Akan.
(6) *Zhan Bu Fa Du*, by Lanmuli.
(7) *Zai Fu Zheng Yi*, by Mata Heli.
(8) *Qiong Li FaJia Shu*, by Haiyati, 7 sections.
(9) *Zhu Ban Suan Fa*, by Hisabiyyah, 8 sections.
(10) *Ji Chi Zhu Jia Li*, 48 sections.
(11) *Xing Zuan*, by Suaal al-Kawakib, 4 sections.
(12) *Zao Hun Yi Xiang Lou*, by Sanadi Alat, 8 sections.
(13) *Zhu Ban Fa Du Zuan Yao*, by Safeina, 12 sections.

Here, "section" equals the concept of "volume" often found in ancient Chinese books. As for items (5), (6), and (7), the number of sections is missing. But according to the annotation "We find altogether 195 sections on astronomy" at the very beginning of the booklist. When 195 is subtracted by the number of the books of the other 10 categories, we can know that the number of the three totals 58.

There have not been specific records on the languages these books are written in. We cannot exclude the possibility that they are Chinese writings, but most probably they are in Persian or Arabic, brought in by Jamal al-Din from Maragha Observatory.

Of all the books mentioned above, their names are paraphrased while the authors' names are sheer transliterations, which make it hard to restore the original text. The situation also accounts for the reason why the progress of identification has been slow. Fang Hao identifies *Si Bo Suan Fa*

Jia Shu with Euclid's reputable masterpiece — *Ji He Yuan Ben* (*Euclid's Elements*), and his opinion seems even more credible with the fact that the two books contain the same number of sections (volumes), i.e., 15 sections (volumes).[23] Some scholars think that Book 4 may be Ptolemy's *Almagest*,[24] with which I am afraid that I cannot agree, since *Zao Si Tian Yi Shi*, as its name shows, deals with the manufacturing of astronomical instruments while *Almagest* is by no means a work dedicated to explaining how to make such instruments. Furthermore, *Almagest* consists of 13 volumes, as opposed to Book 4's "15 sections".

10. Did Islamic astronomy impact Guo Shoujing?

In 1276 A.D., that is, nine years after Jamal al-Din brought in seven instruments, five years after the Huihui Astronomical Observatory was established, and three years after the two observatories of Yuan were put under the administration of Mishujian, Guo Shoujing (1231–1316 A.D.), one of the greatest astronomers in the Chinese history, received an imperial order to design and make a batch of astronomical instruments for Han'er Astronomical Observatory, and he finished the work in 1279 A.D. Much innovation can be found in these newly created instruments, such as the abridged armilla, the scaphe, the square table, and the observing table.[25] Since Guo conducted the work after Jamal al-Din presented seven instruments to Kublai Khan, and his instruments were at that time avant-garde in China, here comes a question — "Was Guo's work impacted by Islamic astronomy?"

As to this question, the answer of most Chinese scholars is "No". They believe that Jamal al-Din's astronomical instruments did not fit into the traditional Chinese astronomy, which is grounded on two reasons. First, the instruments, belonging to ecliptic framework, did not

---

[23] Fang Hao. *History of Exchanges between China and the West*, (1987), p. 579.

[24] Please refer to pp. 214–215 of *The History of Chinese Astronomy* compiled by the research team of Chinese astronomy history.

[25] Please consult *Astronomical Treatise, History of Yuan* for more details of the instruments. As for the most distinguished ones such as the abridged armilla and the scaphe, please refer to pp. 190–194 of *The History of Chinese Astronomy* compiled by the research team of Chinese astronomy history.

conform to the traditional equatorial-centered system in China; second, necessary mathematical knowledge of using these instruments failed to be imported[26] There are some foreign scholars holding the same opinion, like M. Johnson who said, "In 1279 A.D., the designers of astronomical instruments refused to use Muslim techniques which they were so familiar to".[27] Joseph Needham has been rather equivocal on this issue. For example, he thought the evidence was inadequate to determine whether or not Islamic astronomy exerted influence on the making of abridged armilla, but also confirmed the influence based on circumstantial evidence.[28] However, he kept silent as regards what the circumstantial evidence truly was.

In my view, it is hard, at the superficial level, to find any influence of Islamic astronomy on instruments made by Guo Shoujing. Instead, traces of their connection with traditional Chinese astronomical instruments are readily perceivable. I can give a convincing explanation to this.

The core here is that Huihui Astronomical Observatory and Han'er (Beijing) Astronomical Observatory were jointly led by Mishujian, which put astronomers from two sides (astronomers of the Hui nationality and astronomers of the Han nationality) into competition, including Guo Shoujing and Jamal al-Din, two heads of these two observatories. When Guo Shoujing was ordered to make astronomical instruments, he naturally tried to avoid influence from the other side — in a way to show his professional competence, and outshine his opponent. If he allowed the influence of Islamic astronomical instruments in his work, he would have been accused by Hui-nationality astronomers of imitation and inferiority. Then how could Han'er Astronomical Observatory stand out in the competition?

However, if we make an in-depth analysis, it seemed that Guo Shoujing did embrace some influence of Arabic astronomy. Here, I will elaborate on two of Guo's instruments, which attest to my statement.

---

[26] Please refer to p. 202 of *The History of Chinese Astronomy* compiled by the research team of Chinese astronomy history.

[27] M. Johnson. *Art and Scientific Thought*. Fu Shangkui Trans., *et al*. Workers Publishing House, (1988), p. 131.

[28] Joseph Needham. *Science and Civilization in China*. Vol. 4, (1990), p. 481.

Let us first look at the abridged armilla. The innovativeness of the abridged armilla lies in its simplicity — it does not seek to have multiple functions with overlapping rings as the traditional armillary sphere does; instead, it is transformed so that one ring only measures the coordinates of one specific celestial body. The abridged armilla, in fact, consists of two independent instruments on one pedestal, i.e., an equatorial armillary sphere and an altazimuth. The concept of one instrument, one purpose is typical of traditional astronomical instruments in Europe, such as Jamal al-Din's seven instruments and the six instruments created by Jesuit F. Verbiest in the Qing Dynasty under the decree of Emperor Kangxi (they are still intact in the Beijing Ancient Observatory).

Another one is the gnomon. Among the seven instruments offered by Jamal al-Din, one has the same function as a traditional Chinese sundial, namely, a room measuring the shadow cast by the sun to predict solstices, though with higher precision. We can be sure that Guo Shoujing thought little of imitating it, but he did undertake to improve the traditional sundial. He went to Dengfeng, Henan Province, and constructed what in essence is an enormous gnomon. It is known that enormity is a quintessential style found in Arab astronomical instruments.

Of the above two instruments, one represents the European style imported along with Arabic astronomy while the other the Arabic style of Arabic astronomy. Both of them indicate the indirect influence of Islamic astronomy on Guo Shoujing. Of course, this is merely my opinion unless more definite evidence is unearthed.

Chapter 10

Western Astronomy Finds its Way to the Orient in Modern Times (1)

1. Establishment of modern science

In the post-Renaissance times, Copernicus' heliocentric system was put forth and the coming out of De Revolutionibus Orbium Coelestium is seen as the symbol of the rise of modern sciences. More importantly, experimental methods and related concepts were established.

Scientific experimental methods have the following distinctive characteristics:

1. Scientific experimental methods take "objective hypothesis" as the premise, believing that the objective world will not change with people's observation, measurement, or the change of the subjective will of mankind. This premise guided the development of science and technology in the past few centuries and helped succeed at countless achievements. Hence, they had long been considered perfectly justified and unquestionable (it is also seen as the cornerstone of materialism) until the 20th century when a series of new developments in physics started to shake them in the philosophical sense.

2. Scientific experimental methods completely abandon the transcendental, institutive, and mystic methods as well as those old ways the ancient and medieval people adopted to understand the world.

Experiments are prerequisites to ensure the correctness of knowledge as they can be repeated.

3. Scientific experimental methods emphasize the use of "model method" to understand and describe the world, that is, through observation and thinking to construct the model (which can be mathematical formulas, geometric figures, etc.), and then through experiments (new observations in astronomy) to test the conclusion deduced by the model; if the two are more consistent (always consistent only in a certain degree), then the model is deemed successful. Otherwise, it is necessary to modify the model in order to seek further consistency with the experimental results.

Judged as against these three viewpoints, Copernicus' heliocentric system and the De Revolutionibus Orbium Coelestium (1543) are not yet truly scientific experimental methods. For example, Copernicus insists that the movement of celestial bodies must not be contrary to Pythagoras's assertion that the celestial bodies must make circular motions.[1] Another example is that the Copernican system does not have any superiority in describing planetary motion and predicting planetary position when compared with Ptolemy's geocentric system.[2]

The experimental method began to achieve significant results in the 16th century with William Gilbert of Colchester, whose "magnetism" was published in 1600. Many findings in his book came from his various magnetic experiments. Though the famous Francis Bacon did not achieve much in science, he had a great effect on the establishment of scientific experimental methods as a philosopher. Even though his book "The Great Revival of the Academic" emphasizes the induction method, it is actually the first half of the experimental method.

Kepler and Galileo were the first to make great achievements with truly modern scientific experimental methods. Kepler's third law of planetary motion (1609, 1619) and its discovery process are a successful example of the use of model methods in astronomy, in which the traces of

---

[1] S. F. Mason. "A History of the Sciences", translated by Shanghai Foreign Natural Science Philosophy Translation Group, Shanghai People's Publishing House, (1977), p. 119.

[2] Copernicus is satisfied with the discrepancy of not exceeding 10 seconds of arc between the theoretical study and practical observations. Please see A. Berry: A Short History of Astronomy, New York, (1961), p. 89.

ancient thinking creeds are no longer visible. Since then, many new discoveries in astronomy have been made with the abovementioned model methods. Galileo's research in mechanics, though he was a pioneer like Leonardo da Vinci and Simon Stevin, was done successfully with a complete model approach in the real sense. In particular, in the process of using the model method, he was able to ingeniously ignore the influence of secondary factors, thus enabling him to carry out the mathematical processing and ultimately obtain the correct results, paving the way for the extensive use of model approach in scientific research.

In the use of experimental-model methods, deductive reasoning plays a very important part. With the help of appropriate mathematical tools, deductive reasoning can have a great function, so that people think that some experiments only need to be carried out on paper or in mind, just as Galileo said:

By discovering a single factual reason, our knowledge of this fact is enough to make us understand and affirm some other facts without resorting to experimentation. As the present case shows (the fact that the cannon reached the farthest range when fired from the 45-degree elevation) — as evidenced by the fact that the predecessors had discovered the observation, the author could have a firm grasp of the argument to prove that 45-degree elevation ensures the farthest range.[3]

Of course, what is conducted on the paper or in the mind is not an experiment in the real sense. The model method has since then become the most important weapon in developing science and technology. And the great success of Newtonian mechanics and the consequent amazing astronomical achievements brought extreme glory to the model method. What is most noteworthy with the rise of modern science is the establishment and use of model methods, which, in fact, had long been in use in ancient Greek astronomy but had not become a universal method to explore knowledge and it had not yet formed into the modern form we know today.

2. Entry of Jesuits into China and the policy of academic missionaries

At the end of the sixteenth century, Jesuits began to enter China. In 1582, Matteo Ricci (1552–1610) arrived in Macau, China, as the pioneer Jesuit

---

[3] Quoted from *History of Natural Sciences*, p. 145.

missionary mission in China. After many years of efforts, setbacks, and extensive contacts with people from all walks of life in China, Matteo Ricci sought out an effective way to carry out missionary activities in then China — the so-called "academic mission". In 1601, he was allowed to see Emperor Wanli, and was given the consent to stay in the capital, which meant that the Jesuits had been officially accepted by the Chinese upper class and also marked that the "academic mission" policy was taking effect.

Though "academic mission" was often attributed to Matteo Ricci, as a matter of fact this policy owed great debt to the inherent tradition of the Jesuits. The Jesuits always attached importance to education and established many schools. In the 1720s and 1730s, the Jesuits established 19 schools in Naples, 18 in Sicily, and 17 in Venice, Italy.[4] Since the Jesuits were obliged to undergo rigorous education and training, many of them became outstanding scholars. For example, Matteo Ricci learned astronomy from the then famous mathematics and astronomer Clavius, and later became a friend of and colleague to Kepler and Galileo. Another example was Johann Adam Schall von Bell (1592–1666), who was later appointed the head of The Imperial Board of Astronomy, and his teacher C. Grinberger was the successor of Clavius as the professor of Roman College. A third example is Johann Terrenz Schreck (1576–1630), who was a member of the compilation team of the *Chongzhen Almanac*. He was the academician of the Emirates College (Accademia dei Lincei, predecessor of the Italian Academy of Sciences). He was also friends with Kepler and Galileo (also academicians with Accademia dei Lincei). It was the Jesuit's tradition of valuing education that made "academic mission" possible.

We can have a better understanding about the "academic mission" from remarks of Jesuits in China. Here are only two quotations from Matteo Ricci and D. Parrenin (1665–1741), respectively, two figures who lived 150 years after each other.

The conversion of an intellectual is more valuable and influential than that of many conversions.[5] *In order to win the attention of them (mainly*

---

[4]W. V. Bangert, S.J. *A History of the Society of Jesus*, St. Louis, (1986), p. 187.

[5] *"Letters of Matteo Ricci" (Lettres edifiantes et curieuses)*, Translated by Luo Yu, Kuan-hchi Publishing House (Taiwan, 1986), p. 314.

the Chinese intelligentsia), we must obtain their inner trust and win
their respect by making use of their curiosity in knowledge about natural
things. There is nothing better than this to make them easily understand
the holy spirit of Christianity.[6]

If we deliberately want to be critical, we can say that the spread of
scientific and technological expertise of the Jesuits in China only served
as bait. But objectively, the "fish" had bitten the bait after all, which
inevitably exerted an effect on it.

3. Taking a shortcut to reach the top authorities — the earliest attempt of Matteo Ricci

Astronomy in ancient China was not as a natural science, but tinted with
a very strong political color. First of all, astronomy played a role in
politics — during primitive times, it served as the basis for the royal
power, and later it became a symbol of the power.[7] Until mid-Ming
Dynasty, it was only for officials of the imperial Tianxue academies. As
for the army and the people, to practice astronomy in private was seen as
a felony, and this rule lasted for about 2,000 years in Chinese history until
late Ming Dynasty, on the eve of the Jesuits coming into China.[8]

Matteo Ricci entering into the capital coincided with the time the
Ming court was about to modify the *Da Tong Li* (Datong Calendar) as it
was getting increasingly faulty in forecasting astronomical phenomena.
The Ming court had been thinking of modifying the calendar for years.
Knowing this, Matteo Ricci soon made an attempt to participate in revis-
ing the calendar. In his memorial to Emperor Wanli for contributing
Roman local products, he said:

I've grasped the know-hows about making the maps of the universe and
the earth as well as the physique mathématique. I make apparatuses to

---

[6] "The Letters of Jesuits in China", Vol. 24, p. 23; quoted from Jacques Gernet: *"China and Christianity"*, Shanghai Ancient Books Publishing House, (1991), p. 87.

[7] For details, please refer to Chapter 3 of Jiang Xiaoyuan's *Tianxue Zhenyuan*.

[8] Jiang Xiaoyuan: *Tianxue Zhenyuan (The Truth of the Sciences of the Heaven)*, pp. 65–68).

observe the astronomical phenomena to test the sundial in a way that coincides with ancient Chinese methods. If His Majesty wouldn't disdain me and allow me to exercise my humble wisdom to serve in the court, it would be my dream fulfilled.[9]

Though Matteo Ricci failed to win the heart of the Emperor with his self-recommendation, it was the first attempt by the Jesuits to take a "shortcut" to the imperial court. They tried to open up the road to the royal court in the capital with astronomical and calendar expertise.

Matteo Ricci had a very clear understanding of securing a shortcut to the highest authority. As he understood the special status of astronomy in ancient Chinese politics and culture, he strongly urged the Roman authorities to send astronomical Jesuits to China. In his letter to Rome, he said:

It is of great significance and conducive to the missionary work to send a priest or a monk proficient in astronomy to China, for I myself have some knowledge about other technologies, such as watches, globe, geometry. Besides, there are many books of kind for reference. But the Chinese people do not attach importance to them, but show great interest in the orbits and locations of planets and calculating for solar and lunar eclipses, as they are very important to the compiling of the calendar.

... I instruct the Chinese with the world map, watch, globe and other works. In their mind, I'm the greatest mathematician in the world ... so I suggest you send an astronomer to Beijing. I can translate our calendar into Chinese, which is not difficult for me. By doing this, we will be more respected by the Chinese people.[10]

Matteo Ricci meant to strengthen the astronomical force of the Jesuits in China as icing on the cake. In fact, many of the Jesuits in China at that time, including Matteo Ricci, had very high astronomy attainments — their accomplishments in this area fascinated a lot of Chinese officials so much so that they submitted a petition (or a letter) to the throne recommending the Jesuits to participate in the calendar revising. For example, in 1610, Zhou Ziyu, a fifth-rank official with the Imperial Board of Astronomy, rec-

---

[9] Huang Bolu: *Zheng Jiao Feng Bao* (*Imperial Praise of the Orthodox Religion*), the Tz'umu t'ang of Shanghai, (1904), p. 5.

[10] *"Letters of Matteo Ricci"* (*Lettres edifiantes et curieuses*), pp. 301–302.

ommended Diego de Pantoja (1571–1618) and Sabatino de Ursis (1575–1620) to join in the calendar revising work; in 1613 Li Zhizao recommended Diego de Pantoja, Sabatino de Ursis, Manuel Dias (1574–1659), and Niccolo Longobardo (1565–1655). Li's words are quite representative, as seen from the *Li Zhi* (*Calendar*) Book 1 of *Ming Shi* (*History of Ming*):

> *Their views on astronomical calendar touch upon realms that Chinese astronomical sages never treaded on. They not only discussed their physique mathématique, but also the underlying laws. The apparatuses they made to observe the sky and the sun are exquisite.*

These recommendations eventually worked.

4. *Chongzhen Almanac* and the Western astronomical works it is based on

In 1629, The Imperial Board of Astronomy officials made a mistake again when using the traditional methods to calculate the time of solar eclipse, while Xu Guangqi made a correct calculation by using Western astronomical methods. So, Emperor Chongzhen ordered the establishment of an "Almanac Board" headed by Xu Guangqi to compile a new calendar. Xu Guangqi had summoned the Jesuits Niccolo Longobardo, Johann Schreck (1576–1630), Johann Adam Schall von Bell, and Jacobus Rho (1592–1638) to work on the almanac board. From 1629 to 1634, they compiled *Chongzhen Almanac*, known as "European classical astronomical encyclopedia".

"*Chongzhen Almanac*" is voluminous. The theoretical part, taking up one-third of the length, gives an elaborate introduction to the theories and methods of Western classical astronomy, with emphasis on the work of Ptolemy, Copernicus, and Tycho. In general, it does not go beyond the Three Laws of Kepler Planetary Motion, but there is still some more advanced content. The specific calculations and the large number of astronomical tables are based on Tycho Brahe's system. What theories are introduced and what books are used in the "*Chongzhen Almanac*" can be revealed through a textual research.[11] The verified quoted books are as follows:

---

[11] For details, please see Jiang Xiaoyuan: "The Spread and Influence of Western Astronomy in China during the Ming and Qing Dynasties", PhD Thesis of the Chinese Academy

Tycho Brahe
  (*Astronomiae Instauratae Progymnasmata, 1602*)
  (*De Mundi,* 1588, namely Interpreting Comets by the Jesuits in China)
  (*Astronomiae Instauratae Mechanica, 1589*)
  (De Nova Stella, 1573, later reprinted in First Stage)
Ptolemy
  (*Almagest*)
Copernicus
  (*De Revolutionibus, 1543*)
Kepler
  (*Ad Vitellionem Paralipomena, 1604*)
  (*Astronomia Nova, 1609*)
  (*Harmonices Mundi, 1619*)
  (*Epitome Astronomiae Copernicanae, 1618–1621*)
Galileo:
  (*Sidereus Nuntius*, 1610)
Longomontanus
  (Astronomia Danica, 1622, a book by Tycho's student illustrating
  Tycho's theory)
  Purbach & Regiomontanus
  (*Epitoma Almagesti Ptolemaei, 1496*)

The Jesuits carried the above-mentioned 13 books to China despite travelling thousands of miles by land and sea. In addition to these, the compilers of "*Chongzhen Almanac*" cited contents from Latin astronomical works of the 16th and the 17th centuries. Among them, 10 books are still kept in Atlanta Chinese Christian Church North. The latest one was published in 1622, but all were books prior to the beginning of compilation of "*Chongzhen Almanac*".

5. *Chongzhen Almanac* and Copernicus' theory

"*Chongzhen Almanac*" includes a large number of measurement examples that list the calculation schemes based on the models of Ptolemy,

Copernicus, and Tycho,[12] but it does not formally introduce Copernicus cosmic model. In the past, it was generally believed that Copernicus' theory did not enter China until the Jesuit P. Michel Benoist presented Emperor Qianlong *Kun Yu Quan Tu (Atlas Maior)* in 1760. In general, there is nothing wrong with this belief, but as a matter of fact the Jesuit missionaries had not blocked the entry of Copernicus' theory into China before P. Michel Benoist, but introduced some citations from the theory.

Basically, *"Chongzhen Almanac"* borrows Chapter 11 of De Revolutionibus Orbium Coelestium and quoted 17 observations out of the 27 observation records in the book. *"Chongzhen Almanac"* also introduced some important elements of Copernicus heliocentric theory. For example, Wu Wei Li Zhi of the Almanac has the following paragraph about the earth's movement.

That on the ground we see stars move on the left does not necessarily mean that the stars move by themselves, because stars do not rotate and the earth and its atmosphere rotate from the west to the east. One complete rotation means one day and one night. It's like people taking a boat. When they see the trees on the shore, they find the trees are moving, not themselves. This is also the case when people on the ground see stars move west. When the earth rotates, other celestial bodies do not have to. When the earth completes one rotation, the celestial sphere doesn't have to move.

This passage is almost directly translated from Section 8, Chapter 1 of *De Revolutionibus Orbium Coelestium,*[13] and it uses the Earth's rotation to explain the celestial sphere on Sunday as the movement. This is undoubtedly an important part of the Copernican theory.

However, although *"Chongzhen Almanac"* introduces the content, it does not endorse it, believing it "not the normal solution" based on the grounds that "if a person in the sailing boat sees the shore move, doesn't it mean that a person on the shore sees the boat moving, too?" This retort seems tenable. The boat–shore analogy only tells about the relativity of motion, but does not prove the rotation of the earth. In fact, during the

---

[12] Jiang Xiaoyuan: Ptolemian Astronomy Introduced by the Jesuits in Late Ming Dynasty, *Studies in the History of Natural Sciences*, Vol. 8, No. 4 (1989).

[13] Copernicus: *De Revolutionibus*, 1, 8, Great Books of the Western World, *Encyclopedia Britannica*, Vol. 16, (1980), p. 712.

time when *"Chongzhen Almanac"* was produced, no evidence regarding the earth's revolution had ever been found.[14]

6. Waterloo for Chinese Tianxue: Eight clashes with Western astronomy in 10 years

When the *"Chongzhen Almanac"* was compiled, Xu Guangqi, Li Tianjing (who took over the Almanac Board after Xu Guangqi passed away), and others argued a lot with such conservatives as Leng Shouzhong and Wei Wenkui. The former endeavored to defend the superiority of Western astronoy (European mathematical astronomy), while the latter objected to it and upheld the traditional methods of China. When the *"Chongzhen Almanac"* was completed, it should have been promulgated. But because of the fierce opposition from the conservatives, the strife went on for ten years and the promulgation was suspended.

The conservatives opposed the promulgation with the excuse that they doubted the accuracy of the new calendar. However, no matter what the underlying causes for their opposition to Western methods were, they agreed with Xu and Li on judging the pros and cons of their respective astronomical theories with actual observation accuracy (i.e., the degree of agreement between the estimated value of positions of celestial bodies and the actual observed values). The *Li Zhi* (*Calendar*) *of Ming Shi* (*History of Ming*) records eight matches between the two sides, which are truly rare historical records of sciences and culture. These matches followed the same pattern. The two sides calculated the time and location of some astronomical phenomena by using their own methods and then checked these with actual observations to see which side was closer to the actual observations. The matches involved such phenomena as solar eclipses, lunar eclipses, and planetary motion. The following are the years and astronomical phenomena of the eight matches:

1629, solar eclipse.
1631, lunar eclipse.

---

[14] For details of the textual research, please refer to Jiang Xiaoyuan: *The Spread and Influence of Western Astronomy in China During the Ming and Qing Dynasties*, PhD dissertation of CAS (Beijing, 1988), pp. 7–8.

1634, Jupiter movement.

1635, movement of Mercury and Jupiter.

1635, locations of Jupiter, Mars and Moon.

1636, lunar eclipse.

1637, solar eclipse.

1643, solar eclipse.

The results of the eight contests came out at 8:0, with China's traditional astronomical approach meeting Waterloo.[15] Among the eight contests, three occurred before the compilation of the "*Chongzhen Almanac*" and five occurred after the completion of the Almanac. At the seventh, Emperor Chongzhen was well aware of the superiority of the Western methods. After the last match, he made up his mind to promulgate the *Chongzhen Almanac*. But, unfortunately, as the Ming Dynasty was meeting its doomsday, his edict failed to be carried out.

7. From *Chongzhen Almanac* to the Western New Almanac

With ten years of efforts and five years' experience in revising the "*Chongzhen Almanac*", the Jesuits ultimately managed to convince Emperor Chongzhen the superiority of Western astronomical methods. When they were about to open up the road to the highest authorities of the Ming, there came the change of dynasties, which compelled them to seek out new options.

In March 1644, Li Zicheng's army invaded into the capital Beijing, and Emperor Chongzhen hanged himself. But soon, Li was defeated by the allied forces of Wu Sangui and the Qing regime. On May 1, 1644, the Qing army entered the city of Beijing, bringing an end to the Ming Dynasty. Johann Adam Schall von Bell, who was in Beijing at the time, had to make his choice: how could he maintain and carry on his missionary work in China after the regime change? Different from other Jesuits

---

[15] For the explanations of the results of the eight contests, please refer to Jiang Xiaoyuan: On the Superiority of Tycho's Astronomical System-Investigation and Discussion from Three Aspects, *Journal of Dialectics of Nature*, Vol. 11, No. 1 (1989).

who continued to work with the Southern Ming regime. Johann Adam Schall von Bell chose to work together with the Qing regime. Nobody could have imagined that the *"Chongzhen Almanac"* that could not get promulgated even when it had been completed for 10 years became a gift from God to the new regime. The Almanac became a lavish gift to the new regime that badly needed a new calendar to symbolize the mandate of heaven on the change of the regime. Johann Adam Schall von Bell revised and abridged the *Chongzhen Almanac* and presented it to the Qing regime, which was soon adopted by the new regime. The Western New Almanac with the inscription by the Emperor soon got promulgated. Despite being in a precarious state and amid in the raging storms, the Ming regime dedicated so much manpower and material strength in compiling the monumental work of *Chongzhen Almanac*, a task worth commending. Unfortunately, it was not promulgated in a timely manner and eventually became a gift for the new regime.

Johann Adam Schall von Bell, as a result of his contribution of the Almanac and his efforts, was appointed as the head of The Imperial Board of Astronomy, thus beginning the history of about 200 years of the Qing regime's having the Jesuits take charge of the Imperial Board of Astronomy. The ultimate purpose of Johann Adam Schall von Bell's participation in compiling the Almanac was to make use of astronomical knowledge to step into the imperial court of Beijing. He himself was very good at dealing with the court and the nobility. At the end of the Ming Dynasty, he served as the rector of the Jesuit Beijing parish and developed a stron group of believers in the Ming Court. At that time, he managed to convert more than 140 people of the royal family, 50 dowagers, and more than 50 eunuchs into believers. After the Manchu took the throne, Johann Adam Schall von Bell won the favor and trust of Emperor Shunzhi. Shunzhi often called him "mafa", meaning "grandfather" in Manchu language. This was because he cured Empress Dowager Xiaozhuang of her disease, and she later adopted him as her foster father. From this example alone, we can easily get the overall picture of Johann Adam Schall von Bell spreading Christianity in Emperor Shunzhi's court.

Since then, the Imperial Board of Astronomy in Beijing had been the most important stronghold of the Jesuits in China. Besides, as Johann

Adam Schall von Bell had won the Emperor's respect and grace, many concubines, princes, ministers, and others became good friends of Johann, which brought about immeasurable benefits to the missionary work in China.

In his later years Johann Adam Schall von Bell ended up in prison because of the Almanac. He was almost beheaded, but died of illness in the end. He was actually the victim of the last attack launched against the Western astronomy by the conservatives. So far, there have been discussions of this matter by many scholars.[16] Soon after his death, the miscarriage of justice on him was rehabilitated. Ferdinand Verbiest (1623–1688), a Jesuit, succeeded him in the post of heading The Imperial Board of Astronomy. As Emperor Kangxi was keen on astronomy and other Western sciences, he often summoned the Jesuits into the palace to give lectures. Hence, the Jesuits had a sweet time imparting astronomical knowledge to the royals. However, after this period, the privilege and courtesy the Jesuits in Beijing received in the court could never be compared to that received during the Emperor Shunzhi and Emperor Kangxi periods. But the official status of the Western astronomical theory and methods made by imperial order remained till the end of the Qing Dynasty.[17] "Western Methods" became a compulsory subject for all students of astronomy during the Qing Dynasty.

Astronomy enjoyed a sacred position in ancient Chinese society. In such a special discipline, using Western methods and posting Westerners in the court has great symbolic and exemplary significance. It can be said that led by astronomy Western ideas, philosophies and methods related to science and technology gained their entry into China during the Ming and Qing Dynasties, some of which were accepted and adopted, exerting a very profound impact.

---

[16] For relatively new discussion, please see Huang Yinong: Struggle on Choice of Date and Kangxi's Calendar Imprisonment, Tsing Hua Journal of Chinese Studies (Taiwan, China), New Vol. 21.

[17] To a large extent Tycho's astronomical system kept its official status, which remained unchanged even with the publishing of *Li Xiang Kao Cheng* (1722) and *Li Xiang Kao Cheng Hou Bian* (1742).

8. Johann Adam Schall von Bell's Adaptations to *Chongzhen Almanac*

Under the supervision of Xu Guangqi and Li Tianjing, the "*Chongzhen Almanac*" was sent in five batches to Emperor Chongzhen for reading, totaling 44 categories and 137 volumes. Though the "*Chongzhen Almanac*" failed to be issued for enforcement in the late Ming, there were in fact editions that had come out, which were called "Ming Editions". When the Qing army entered Beijing, Johann Adam Schall von Bell had the printing plates of the Almanac with him, which he called "Small Plates". The *Xi Yang Xin Fa Li Shu* (*Western New Almanac*) revised by Johann Adam Schall von Bell, were printed many times, and there were quite a lot of editions. Among them there are two representative ones; one was printed in the second year of the Shunzhi Period (Shunzhi edition for short), which is stored at Beijing Palace Museum, while the other is at the Library of Congress in the United States, and this version, according to Wang Zhongmin, was printed during the Kangxi Period. Wang judged it according to Johann Adam Schall von Bell's conferred title "Tong-Xuan Tutor". As Xuan was taboo to Emperor Kangxi, to avoid it Xuan was changed into Wei. So Wang predicated this as the edition printed during the Kangxi Period. Therefore, it is called Kangxi edition. Johann Adam Schall von Bell amended the "*Chongzhen Almanac*" mainly in the following two ways.

One was by deletion. The "Western New Almanac" Shunzhi Edition falls into only 28 categories while the Kangxi edition has 27 categories with 90 volumes. Deletions were made mainly in various astronomical tables, but the theoretical part of the *Chongzhen Almanac* remained unchanged, including information on the solar movement, the lunar movement, star movement, eclipse movement, and planet movement.

The second was addition. The added articles in the "Western New Almanac" were mostly short, and were mostly written by Johann Adam Schall von Bell himself. There were also articles by others, such as J. G. Aleni's *ki-ho yao-fa* (*Compendium of Euclid's Elements*) (dictated by J. Aleni, and taken down by Qu Shigu (瞿式谷), as well as old works of almanac authorities such as *Theory of the Armillary Sphere* (written by Johann Adam Schall von Bell and revised by Giacomo Rho). As the

Shunzhi Edition and the Kangxi Edition of the *Western New Almanac* are rare books, the additions in the *Chongzhen Almanac* are picked out and listed in the following table (the asterisks in Column 3 and Column 4 indicate the works are included).

Name of Work	No. of Volumes	Shunzhi Edition	Kangxi Edition
Li-Shu (Calender Commentary)	2	*	
Zhi-Li-Yuan-Qi (Why Compile the Almanac)	8	*	*
Xin-Li-Xiao-Huo (Commentary on Compiling the New Almanac)	1	*	
Xin-Fa-Li-Yin (Guide of New Almanac)	1	*	*
Ce-Shi-Lve (Outline of Measuring Eclipses)	2	*	*
Xue-Li-Xiao-Bian (Arguments on Almanac Studies)	1	*	*
Yuan-Jing-Shuo (Introduction to Telescope)	1	*	*
Ki-ho yao-fa (Compendium of Euclid's Elements)	4	*	*
Hun-Tian-Yi-Shuo (Introduction to the Armillary Sphere)	5	*	*
Chou-Suan (Arithmetic)	1	*	*
Huang-Chi-Zheng-Qiu (Spherical Astronomy on the Ecliptic Equator)	2	*	
Li-Fa-Xi-Chuan (Chinese Calendar Going to the West)	1		*
Xin-Fa-Biao-Yi (Improvements Made in the New Almanac)	2		*

As far as the objective effect is concerned, the amendments by Johann Adam Schall von Bell made the *Western New Almanac* more compact and complete than the "*Chongzhen Almanac*". At the same time, it cannot be denied that the additions of nearly ten writings by Johann Adam Schall von Bell leave readers the impression that Johann Adam Schall von Bell contributed a lot to monumental work, as bigwigs could never scrutinize the whole work and so just went through the table of contents at most. Although Johann Adam Schall von Bell was one of the two most important compilers of the *Chongzhen Almanac*, he amended the *Chongzhen Almanac* before presenting it to the Qing government as a gift, for which we have every reason to say that he had the motive of appearing important to the new regime by revising and presenting the Almanac.

9. Western astronomy during the rise and fall of the Ming Dynasty

Considering the historical background of Western astronomy finding its way to China at the turn of the Ming and Qing dynasties, there should be one thing worth noting, the so-called "Thought of Pragmatism" in the late Ming Dynasty, which is a term used by modern people. As the Ming scholars lived in peace and extravagance for a long time, obsessed with various material and spiritual enjoyment, they did not care much about enriching the country and increasing its military force or minding pragmatic matters, for which Xu Guangqi criticized them as "despising worldly affairs". Modernists often blame this phenomenon on the prevalence of the "philosophy of the mind" advocated by Lu Jiuyuan and Wang Yangming. But this is an arbitrary argument that cannot be made easily and is definitely not within the scope of this book.

Even seen from more positive aspects, Ming scholars' obsession of moral and spiritual aspects did not help dissipate the internal and external troubles the court faced at the end of the Ming Dynasty. Even though the parties like Donglin and Fushe were worth lauding as they dared to denounce the evil forces, Mr. Liang Qichao (1873–1929) ridiculed it as "a fight between disciplines of Wang Yangming and those of Wei Zhongxian, two groups of Confucian scholars". However, the ridicule was of a pedantic nature and did not help at all. Yan Yuan (Xi Zhai) once uttered the well-known line "Confucian scholars like discussing the philosophy of the mind and will sacrifice their lives in case of emergency to repay the

Emperor", which is a typical mindset of Ming scholars who believed "talking about the philosophy of the mind" was a contribution to the society. What a ridiculous mindset it was!

On the other hand, when the Ming Dynasty in the late years was caught in the dilemma of internal and external troubles, some scholars realized that the empty talk of the mind of philosophy could not help at all in saving the doomed dynasty. Thus, they called upon people to do practical work and practice pragmatic learning, and they earnestly practiced what they advocated. Xu Guangqi was a representative figure among them in this regard. Though he was determined to save the nation, unfortunately he was powerless to save the desperate situation. He died with his dream unfulfilled.[18]

When the Manchu troops invaded the Ming territory with tremendous cavalry, the Ming Dynasty fell apart instantly. Only when the Ming scholars who were used to the empty talk of the philosophy of the mind were conquered by Manchus did they wake up from the dream. Some of them began to reflect deeply. The so-called "pragmatic learning" at the end of the Ming Dynasty rose out of the collapse of the dynasty. The representative figures were mostly adherents of the former Ming Dynasty. Liang Qichao (1873–1929) had the following account of this school of thought:

> *Although these scholars grew in the prevailing climate of the Yangming School, for the sudden changes of the social situation, their minds changed like silkworms and started to breed new lives. They thought the fall of the Ming Dynasty was the greatest shame on them. Hence, they abandoned the empty talk of the philosophy of the mind and focused on humanistic pragmatism. They engaged themselves in academic studies not for the sake of the academics but for politics. Many of them wasted half of the lifetime on dire politics as they originally intended to prepare themselves for the new political development. Having lost all hope for politics, they turned to sheer academic life.[19]*

---

[18] Liang Qichao: *"History of China's Academics in Recent 300 Years"*, included in *"Liang Qichao Talks on Academic History of the Qing Dynasty (Two Categories)"*, Fudan University Press, (1985), p. 94.

[19] Liang Qichao: *"History of China's Academics in Recent 300 Years"*, p. 106.

The most famous ones of these scholars are Gu Yanwu, Huang Zongxi, Wang Fuzhi, and Zhu Shunshui. The first three are usually referred to as Three Masters, who were deemed spiritual leaders of intellectuals at the turn from the Ming to the Qing. They were much respected as they chose not to cooperate with the Qing regime and maintain the identity as adherents of the former Ming Dynasty. Besides, they were masters of humanistic pragmatism.

At the turn of the Ming to the Qing dynasties, some scholars were particular about humanistic pragmatics (Modern people prefer to address them as masters of humanistic pragmatism just because they were engaged in science and technology), such as Gu Yanwu, Huang Zongxi, Wang Fuzhi, and Fang Yizhi. Sometimes they were called by modern scholars as scholars of enlightenment. However, this argument could easily trigger controversy, which shall not be discussed here. But, as a matter of fact, these scholars and their work had paved the way for Chinese science ideology to move on to a new phase.

Chapter 11

Western Astronomy Finds its Way to the Orient in Modern Times (2)

1. Western concept of a spherical Earth

There have been many records on the spread of the concept of a spherical Earth in China. However, I would like to explore the concept with the following aspects, which might have been ignored in previous studies, as the concept of a spherical Earth is an indispensable part of the models of the universe.

From the perspectives of astronomical theories and the relations of the models of the universe, the concept of a spherical Earth, which found its way to China in the Ming dynasty, has the following two highlights:

1. The earth is spherical.
2. The earth is tiny compared with the universe.

The former statement has been precisely discussed in previous studies while the latter needs further illustrating. In the late Ming dynasty, Jesuit missionaries presented evidences showing the earth is round from every aspect, but few of these have directly made it clear that the earth is much smaller compared with the universe. Actually, Western astronomers had always held that the earth was much smaller, so there was no need for further arguments. For instance, "*Wu Wei Li Zhi*" of *Chongzhen Almanac* discussed the distances between earth and five major planets and provided the following figures:

Saturn: distance from earth 10550 earth radius
Jupiter: distance from earth 3990 earth radius
Mars: distance from earth 1745 earth radius

These figures are wrong according to modern astronomy, but we can still see the relative sizes in Western models of the universe, which indicated that the earth was quite small. Meanwhile, it was mentioned in the third chapter of "Star Movement" in *Chongzhen Almanac* that "Hengxingtian" (the sky of stars) was 14,000 earth radiuses from us. Examples of this kind are too numerous to mention.

The reason why the second point of the concept is so important is that many basic theories of azimuth astronomy are built on the idea that the earth is much smaller than "tian" (the universe). The Westerners have been sure about this since the time of the ancient Greeks. Only in a few circumstances, when there is a horizontal parallax for instance, do we need to take the radius of the earth into consideration. Most of the time, the earth can be seen as a point. The same is true in modern astronomy.

Besides, it is controversial as to whether there is such concept in ancient China. As we saw in Chapter 8 of the book, even if there was a similar idea, it had a big difference from the Western one. The image of the universe widely accepted by astronomers in ancient China shows that the radius of the earth is half of that of "tian" (the universe), which indicates that they are of the same order of magnitude and in no case can the earth be seen as a spot. Since late Ming Dynasty, however, scholars have been arguing that there had been a concept of a spherical earth in ancient China, ignoring the huge difference above. And many modern studies also made similar mistakes.

2. Chinese scholars' rejection of the spherical earth theory

On the reaction of the Chinese to the Western concept of a spherical earth, previous works tended to focus on how people accepted or agreed to it.

Some scholars, for instance, believed that there were mainly two schools in Ming and Qing Dynasties, one which believed there had never been an idea of a spherical earth in China while the other argued such an idea had existed in China since ancient times. Both schools accepted the Western concept of a spherical earth. Nevertheless, there was actually another school that rejected the idea of a spherical earth. We will see from the following analysis that several renowned astronomers who were against it.

In the past few years, it was found that the works of Song Yingxing, a well-known scientist in the late Ming Dynasty, could be divided into four categories and were written in the Chongzhen Period (1628–1644). One of the four is *Tan Tian* (*On Universe*),[1] which has in it the following statement[2]:

> *"The Westerners believe that the earth is round, floating in the air, and everything is attached to it like ants. The feet of the people in Mabazuo are opposite to those in China (which means they are on two ends of the earth). Such comments on the shape of heavenly bodies are totally nonsense, just like the statements in xuanyeshuo (late-night discussion, an astronomical theory in ancient China that believes the sky has no particular shape and the heavenly bodies follow the movement of qi) and Zhou Bi Suan Jing written during the Western Zhou".*

Apparently, the Western theory Song referred to came from Matteo Ricci.[3,4] I have to say that the astronomical theory held by Song is extremely primitive. He even believed that the sun was immaterial and

---

[1] See Lin Jinshui. "The impact and significance of the concept of a spherical world brought by Matteo Ricci". *Journal of Literature, History & Philosophy*, No. 5, (1985).

[2] In 1894, J. F. Herschel, a British astronomer, completed his book *The Outlines of Astronomy*, which was a hit in the Western world. Ten years later, it was translated by Li Shanlan and Alexander Wylie into Chinese and was also named *Tan Tian*, which was not the same book mentioned in this Chapter.

[3] Song Yingxing. *Yeyi·Lunqi·Tantian·Silianshi*. Shanghai People's Publishing House, (1976), p. 101.

[4] Statements about "Mabazuo opposite to zhiren (people whose feet are of a shape opposite to normal people)" can be seen in the captions of Matteo Ricci's world map. We have no idea where Mabazuo is but it is believed to be in today's Argentina. Refer to Cao Wanru: Extent Studies on Matteo Ricci's World Map. Cultural Relics, (1983), No. 12.

that the sunrise and sunset are just the gathering and dispersal of "yang qi" (exuberant air).[5]

Wang Fuzhi, one of the three renowned Chinese philosophers (Gu Yanwu, Huang Zongxi, and Wang Fuzhi) of the late Ming and early Qing dynasties, was strongly against the concept of a spherical earth. He neither agreed to the idea of a spherical earth nor believed it had existed for a long time in the West:

> *"After he came to China, Matteo Ricci heard the Chinese saying the earth was like a slingshot ball (which actually means some place has a very tiny space). He tried to appear clever and said the earth was as round as a ball, taking the expression out of context. However, it is known to all that the ground can be rugged so it couldn't have a shape. Yin and Yang were considered as a ball by him. What a fool!"[6]*

For lack of the knowledge of the latitude and longitude in astronomy, Wang Fuzhi criticized the idea that 125 kilometers on the earth equaled one degree in the sky. He insisted that the shape or size of the earth could never be measured —

> *"Living on the ground himself, Matteo Ricci saw as far as others. With the help of a telescope, however, he insisted that the earth was 4,500 kilometers wide. In the past century, few has realized how crazy he was. Alas!"[7]*

Obviously, this is a criticism made by a layman, and an emotional one at that. It can be inferred from Wang's works that he had more or less read about the Chongzhen Alamance. In the appendix to *Si Wen Lu*, for instance, he quoted from "*Xin Fa Da Lue*" several times, but it seems that he was not convinced by it.

Yang Guangxian was also an opponent of the concept of a spherical earth, which is quite understandable since he had a reputation for

---

[5] Song Yingxing. *Yeyi·Lunqi·Tantian·Silianshi*, (1976), pp. 101–103.

[6] Wang Fuzhi. *Si Wen Lu·Si Jie*. Zhonghua Book Company, (1956), p. 63.

[7] Wang Fuzhi. *Si Wen Lu·Si Jie*. p. 63.

criticizing missionaries. His argument was different from that of Song and Wang. Yang said:

> *The root of this absurd idea lies in the map of their religion, which says the earth is round.*[8]

He believed that a spherical world played an important part in Western astronomy, which is all right. However, he could not accept the idea about zhiren (people whose feet are opposite to ours) and attacked it.

> *They didn't even consider that people on the other end of the earth would have to stand on their head. Such comments are just lies made up by fools on the spur of the moment. If anyone who is not insane think about it based on a reasonable theory, he will find it so ridiculous that he will spit his drink.*[9]

However, the "reasonable theory" he referred to was simply the ancient theory of "Tian Yuan Di Fang" (round sky and square earth):

> *The sky is round and the earth square, which has already been illustrated by our ancestors in detail. The heavy ones came down and became the earth, square and still.*[10]

This is even more unscientific than Wang Fuzhi's comments. The three scholars' rejection of a spherical earth lies in the fact that their intellectual architecture is totally different from that of the Western astronomy. The two sides could not actually talk to each other as they had nothing in common in judgment or expression. In this case, they were just adding their own arguments without understanding each other.

On the other hand, those who accepted Western astronomy had built some intelligent architecture similar to the Western one. It is significant that the astronomers of the Ming and Qing dynasties, who are widely recognized by the modern academic world, all accepted the concept of a spherical earth, Xu Guangqi, Li Tianjing, Wang Xichan, Mei Wending,

---

[8] Wang Guangxian. *Bu De Yi*. Vol. 2. copy of, (1929), p. 63.
[9] Wang Guangxian. *Bu De Yi*. Vol. 2. p. 67.
[10] Wang Guangxian. *Bu De Yi*. Vol. 2. pp. 68–69.

and Jiang Yong included. One major reason for this may be that the evidences the Western astronomy provided (including that if one keeps heading north, he would observe an increase in the altitude of Polaris, that the way masts of ships are visible when the hulls are still out of sight and that the shadow of the earth on the moon during an eclipse is round) was acceptable to scholars with astronomical prowess. On the contrary, those who attacked the idea of a spherical earth were mainly those who lacked astronomical prowess, such as Song, Wang, and Yang.

3. Arguments of Zhang Yongjing and Mei Wending

Here is an interesting anecdote that can tell us a little more about Chinese scholars' attitude towards the concept of a spherical earth at that time.

Zhang Yongjing, styled Jian'an, was born in Xiushui, Zhejiang. "He studied hard and wrote well. Devoted to calendrical science, he wrote *Ding Li Yu Heng*", a book focused on traditional Chinese calendrical science. He presented his book to Pan Lei, who told him that calendrical science was too profound to understand and one could not hold on to one's own views (indicating that the traditional calendrical science Zhang had studied was outdated and he should learn about the Western astronomy that spread to China in the late Ming Dynasty) and suggested him to visit Mei Wending to have a better understanding of this field. Then, Zhang Yongjing traveled a long way to visit Mei, who was so pleased that he invited Zhang to stay with as a guest for over a year, during which time Zhang and Mei exchanged what they had learned about astronomy. After that, Zhang Yongjing wrote a book called *Xuan Cheng You Xue Ji* (On My Study Tour in Xuancheng) to record their productive discussions. Unfortunately, the original copy of the book was destroyed in the Cultural Revolution (1967–1977).[11] The preface written by Pan Lei, however, is preserved, which reads:

> *Zhang Yongjing returned from Xuancheng after over a year and told me:*
> *"I got to the bottom of calendrical science through this trip. Now, I've*
> *come to realize that both Western and Chinese calendrical science have*

---

[11] Bai Shangshu. *A Track Record of Xuan Cheng You Xue Ji*. International Seminar on the 350th Anniversary of the Birth of Mei Wending (Xuancheng, Hefei China, 1988).

their advantages and disadvantages. Together, they can make a good combination. This is the very benefit of a discussion with a friend". Later, he showed me another essay of his and said: "I had hundreds of arguments with the fellows of Wuyan (a school founded by Mei Wending). We have reached many agreements, which don't need further discussion. When it comes to the Western concept of a spherical world, I can never agree. I have discussed with Kun Di and Wang Qiaonian over and over again, with the notes of our discussion amounts to thirty or forty thousand words. This essay is a record of the discussions.[12]

We can infer that *Xuan Cheng You Xue Ji* is mainly about their discussions on whether the earth is round. It is noteworthy that even though Mei Wending, who was a master of both Chinese and Western calendrical studies, and other scholars debated with Zhang Yongjing, they failed to persuade him to accept the idea of a spherical earth, which can shows us how difficult it was for some Chinese scholars to accept this concept.

4. The impact brought about by a world map

The world map made and printed by Matteo Ricci played an important role in transforming the traditional outlook of the Chinese about the universe.

It was in the living room of his home in Zhaoqing, Guangzhou, that he showed the map to Chinese people for the first time. Novel and exotic, the map brought a shock to its beholders and inspired Chinese scholars to explore new fields. Matteo Ricci left a detailed record about it:

On the wall of our living room was a Yu Di Quan Tu (the Great Universal Geographic Map), which was marked in a foreign language. When told this was a map of the entire world, senior Chinese scholars were shocked. They thought China was the whole world and called it Tian Xia (under the sky). They could hardly believe it when hearing that China was only part of the eastern world. So they wanted to know more about it for a better judgment.[13]

---

[12] The complete preface is included in *A Track Record of Xuan Cheng You Xue Ji*.

[13] Matteo Ricci. *De Christiana Expeditione apud Sinas Suscepta ab Societate Iesu*, translated by Liu Junyu and Wang Yuzhuan, Taipei: Guangqi Press, (1986), p. 146. There's also

Matteo Ricci added that:

With an unrealistic and arrogant outlook of the universe, the Chinese believe there's no other country that is better than China. They think China boasts a vast territory, a profound culture as a land of civilization, regarding other countries as uncivilized and barbarous. To them, no other countries have emperors, dynasties or cultures like China do. (......) When shown the map of the world, the uneducated would even laugh at it while the educated would respond differently, especially after they were told about latitude, meridian, equator and tropic lines.[14]

Exaggerated as his remarks may sound, it is in general agreement with the facts. Actually, many of the educated perosns contributed a lot to the spread of new astronomical concepts. They translated Matteo Ricci's descriptions in the map into Chinese and printed it to spread the new knowledge. The educated Chinese showed great interest in his map, which can be seen from the following facts: from 1584 to 1608, 12 versions of the world map appeared across China. These are the results of Hong Weilian's research[15]:

Shan Hai Yu Di Tu (*the Geographic Map*), Zhaoqing, 1584.
Shi Jie Tu Zhi (*World Map*), Nanchang, 1595.
Shan Hai Yu Di Tu (*the Geographic Map*), Suzhou, 1995, 1998 (stonework).
Shi Jie Tu Zhi (*World Map*), Nanchang, 1596.
Shi Jie Tu Zhi (*World Map*), Nanchang, 1596.
Shan Hai Yu Quan Di Tu (*the Great Universal Geographic Map*), Nanjing, 1600.
Yu Di Quan Tu (*the Universal Geographic Map*), Beijing, 1601.
Kun Yu Wan Guo Quan Tu (*the Geographic Map of All Countries*), Beijing, 1602.

---

a Chinese version on the mainland translated from English (English from Latin, and Latin from Italian). Taiwan version was translated directly from the original version, which is in Italian.

[14] *De Christiana Expeditione apud Sinas Suscepta ab Societate Iesu*, p. 147.

[15] Hong Weilian. Matteo Ricci's World Map, *Collection of Hong's Work*, Zhonghua Book Company, (1981).

Kun Yu Wan Guo Quan Tu 2(*the Geographic Map of All Countries version 2*), Beijing, 1602.
Yu Di Quan Di Tu (*the Great Universal Geographic Map*), Guizhou, 1604.
World Map, Beijing, 1606.
Kun Yu Wan Guo Quan Tu (*the Universal Geographic Map of All Countries*), Beijing, 1608, hand drawn

The spread of the Western concept of a spherical Earth is inextricably intertwined with that of world map. This novel concept made the Chinese realize they were not the center of the world. It was the enlightenment without which the Chinese would never make it to the modern age.

That said, it took a long time for most Chinese to develop a common sense of "earth", "world", and "five continents". Before floods of Chinese people could travel abroad, conservative scholar-bureaucrats had a hard time accepting the fact that there were other countries and civilizations. Ultraconservatives would make vituperative comments about the new concepts, and even the moderates would doubt the truth of it. In Volume 3 "Matteo Ricci's attempt to entrap people through heresy" of *Sheng Chao Po Xie Ji* (an anthology of monks and laymen's essays compiled in late Ming Dynasty to protest Catholicism), a statement of Wei is as follows:

Matteo Ricci has been trying to fool the public with his heresy and many scholar-bureaucrats should fall for his absurd remarks about the world. The Atlas Maior he authored is totally baseless as no one has ever seen or set foot on the places on the map. What he drew is like the ghosts drawn by a painter, which can never prove to be true. Unbearably, China was drawn in the northwest of the map......How can China be so small and lie in the north of the world? This is outrageous.

In fact, having already considered the pride of the Chinese, Ricci had tried his best to draw China as close to the center of the map as possible. This Wei, who was the governor of Hunan and Hubei Province, however, acted like those who were described as the uneducated by Ricci. Around 150 years later, when commenting on Giulio Aleni's *Zhi Fang Wai Ji* (*Record of Foreign Lands*) and Ferdinand Verbiest's *Kun Yu Quan Tu* (*Full Map of the World*), officials who took part in the compilation of *Si*

Ku Quan Shu stated that "what they said is exaggerated and baseless. However, it's a funny old world. It shall be recorded in the collection, which can add some tales to it......" or something on the lines of "it is possible that when they came to China and read about ancient Chinese stories, they adapted those stories into something novel and groundless. It wouldn't hurt, however, to add the tales to our collection". At least part of the Chinese had been exposed to a bigger world. Emperor Kangxi (1654–1722) once ordered members of the Society of Jesuits to make maps of China with modern cartography and measurement, which is a case in point. It was a progressive move, domestically and internationally. Just as Fanghao (1910–1980) put it:

> *Few European countries had conducted or completed their nationwide*
> *surveys by the 17th and 18th centuries while China made it, which was*
> *a milestone of the cooperation between Chinese and Western academic*
> *societies.*[16]

Emperor Kangxi himself was really open minded for that time. It is a shame that he failed to take advantage of his power and open the eyes of more Chinese people, which will be discussed later in this book.

5. Aristotelian Cosmology and Joseph Needham's misunderstanding

Major western models of the universe all found their way to China in the late Ming dynasty. The interpretations of, the adaptations to, and the debates about the models had great influences on the ideologies of Chinese scholars, which will be further illustrated in the following sections.

A detailed introduction to Aristotelian Cosmology is given in Ricci's Chinese work *Qian Kun Ti Yi*. Aristotelian universe, as recorded in the book, is a spherical cosmos. The earth stays still, surrounded by concentric celestial spheres, where the moon, Mercury, Venus, the sun, Mars, Jupiter, Saturn, and the stars exist. The ninth sphere is called "the

---

[16] Fang Hao. *Zhong Xi Jiao Tong Shi* (*History of Chinese and Western Transportation*), Yuelu Publishing House, (1978), p. 868.

Primum Mobile". "These nine spheres wrap each other like onion peels, firm and steady. Fixed in them like knots in wooden boards, the stars only move with the spheres, which are bright and colorless like crystals, letting the lights travel through easily". Most of these statements are quoted from *On the Heavens*, Aristotle's chief cosmological treatise. (Only "the Primum Mobile" might have been added to his cosmology by later scholars.)

Before long, *Tian Wen Lüe (Explicatio Sphaerae Coelestis)* by Manuel Dias provided a similar cosmological idea except that there are 12 spheres in it. "The outmost is the 12th sphere, where the gods live. It is heaven, eternally still and extremely large. And the 11th is the sphere of Primum Mobile......"

The abovementioned two books are science books rather than astronomical treatises. The Aristotelian universe is also called "crystalline sphere", which spread to China earlier than other models but had the least influence on Chinese scholars. This is mostly because of the attitude of *the Chongzhen Almanac* towards it. In *"Wu Wei Li Zhi"* of *Chongzhen Almanac*, the differences between Aristotelian universe and Tycho Brahe's were discussed, and the latter was favored.

> *Q: Scholars of the old times believed that heavenly bodies were steady and luminous. Now, however, the obit of Mars and that of the sun intersect with each other. Isn't it obviously contrary to what the ancient scholars said? A: since the ancient times, the key to astronomical observation is following the changes of the heaven and drawing conclusions completely based on the observation figures. This is the right way. Otherwise, how are we supposed to stick to what the ancestors said and run against heaven? It entails us to get rid of the outdated and embrace the new, which is not being clever or violating rules.*

With a high status in Ancient Chinese Science, which has been mentioned in the previous chapter, *Chongzhen Almanac* had a far more significant impact on Chinese astronomy than the works of Matteo Ricci and Manuel Dias. It is not a surprise that Aristotelian Cosmology was not widely accepted as it was denied in the book. As a matter of fact, we are not able to find any Chinese astronomers who were in favor of Aristotelian universe, from the late Ming dynasty to Qing Dynasty. Even if there were some scholars mentioning the crystalline sphere, they were just quoting from *Chongzhen Almanac.*

In terms of the universe models in the Ming and Qing dynasties, Joseph Needham had some misconceptions, which left a lasting influence. Here is a statement of his that is frequently quoted in the fields of history of science, history of philosophy, and even history of China.

> The world map brought by the Jesuits was based on Ptolemy's and Aristotle's geocentric theories, which believed that the universe was a spherical crystal system with the earth as its center. ......Concerning the structure of the universe, missionaries attempted to apply a wrong model (crystalline sphere) to a generally right one (a system from xuanye theory that believes the stars are floating in the sky).[17]

The statement has the following defects. To begin with, the crystalline sphere actually had nothing to do with Ptolemy, who had never been in favor of the model.[18] The truth is, in the late Middle Ages, T. Aquinas quoted from Ptolemy's works to prove that the universe was geocentric and the earth was still when he fully combined Aristotelian Cosmology with Christian theology. Thus, it would be inappropriate to ascribe crystalline sphere to Ptolemy.

Besides, Joseph Needham failed to take into consideration *Chongzhen Almanac*'s rejection of the crystalline sphere and the book's extensive and decisive impact on astronomy in the Qing Dynasty. Instead, he made an argument based on the popular science readings written by Matteo Ricci and Manuel Dias, which led to the wrong conclusion. Meanwhile, other astronomers from the Society of Jesus and Chinese astronomers were all against the model. In this case, Needham's statement that Jesuits imposed the crystalline sphere on the Chinese people is completely without foundation.

6. Ptolemy's model

Made of a team of four led by Xu Guangqi, including missionary Johann Adam Schall von Bell in 1634, *Chongzhen Almanac* is a magnum opus

---

[17] Joseph Needham. *Science and Civilization in China*, (1990), p. 643 and p. 646.

[18] Refer to Jiang Xiaoyuan. "Crystalline sphere in the history of astronomy", *Acta Astronomica Sinica*, Vol. 28, No. 4 (1987).

that provides a systematic introduction to the classical astronomy of the west. It described Ptolemy's model in terms of the movement of planets, which can be seen in "zhou tian ge yao xu ci di yi" (the order of planet movement) of "Wu Wei Li Zhi" in the book. And the "qi zheng xu ci gu tu" (old picture of the movement of planets) is actually a diagram of Ptolemy's model.

Although in Ptolemy's model, the earth is also still and the distances between stars are inconsistent with those in Aristotelian universe, there are no real spheres in Ptolemy's model and the "celestial spheres" are just geometrical demonstrations of the motions of heavenly bodies.[19]

In addition, mathematical descriptions of observations are based on a predicted motion of a small sphere added to the model while Aristotle's cosmological model is based on the difference in the rotation speeds and the angles of spheres.[20] *Chongzhen Almanac* gives plenty of calculation examples based on how Ptolemy's model maps out the position and motions of the heavenly bodies.[21]

7. The Tychonic cosmological model and its official status

With Tychonic cosmological model as the theoretical basis of *Chongzhen Almanac*, all astronomic diagrams in the book are based on this model. Description of qi zheng xu ci xin tu" in (new picture of the movement of planets) "zhou tian ge yao ci xu di yi" (the order of planet movement) of "Wu Wei Li Zhi" in the book reads that:

> *The earth is in the middle, the heart of the spheres formed by the sun, the moon and the stars. The sun itself is the centre of two small spheres of Venus and mercury. The sphere of Mars, which is larger, intersecting with that of the sun. Outside are the largest two, spheres of Jupiter and Saturn.*

---

[19] Ptolemy: *Almagest*, 1X2, Great Books of the Western World, *Encyclopedia Britannica*, Vol. 16, (1980), p. 270.

[20] Refer to Jiang Xiaoyuan. "Crystalline sphere in the history of astronomy".

[21] Refer to Jiang Xiaoyuan. "Ptolemy's model brought by missionaries in the late Ming Dynasty, *Studies in The History of Natural Sciences*, Vol. 4, No. 4 (1989).

That is to say, the earth is surrounded by concentric celestial spheres of the sun, the moon, and the star while the five planets revolve around the sun. This is basically a middle way between Ptolemy's geocentric model and Copernicus' heliocentric system. *Chongzhen Almanac* also high-lighted that the spheres in the model are not entities.

The spheres can intersect with each other, which means they are not celestial bodies that really exist.

Calculations based on how Ptolemy's model maps out the movement of heavenly bodies can be found everywhere in the book. Edited by Johann Adam Schall von Bell, *Chongzhen Almanac* was then presented to the imperial court of the Qing Dynasty. Renamed as *Xi Yang Xin Fa Li Shu* (*Western New Almanac*), the book was issued in the second year of the reign of Emperor Shunzhi (1645) of the Qing Dynasty and became the official astronomic treatise of the dynasty. In the 61st year of the reign of Emperor Kangxi (1722), a team of scholars was assembled to compose *Li Xiang Kao Cheng*, which was a revised version of *xi yang xin fa li shu*, with a few examples and figures replaced. *Li Xiang Kao Cheng*, however, was also based on Tychonic cosmological model, and many data in the book were from the model. The Tychonic model remained the dominant astronomic theory as *Li Xiang Kao Cheng* was "written by the imperial court".

By the seventh year of the reign of Emperor Qianlong (1742), *Li Xiang Kao Cheng Hou Bian* had been composed, in which the motions of the sun and the moon were calculated based on Kepler's first law and second law. This could have meant a break with the Tychonic model but the composers cleverly switched the positions of the sun and the earth, which made no difference to the mathematical result. In this case, the system was still geocentric. If such a system was applied to the movement of planets, it would not add up. Nevertheless, *Li Xiang Kao Cheng Hou Bian* illustrated only the motions of the sun and the moon and the intersection of spheres without mentioning the motion of plan-ets. In this way, possible paradoxes were perfectly avoided. Afterwards, Li Xiang Kao Cheng Hou Bian and Li Xiang Kao Cheng were released as one, which made the Tychonic model retain its official status, at least theoretically.

8. Wang Xichan's and Mei Wending's adaptations of the Tychonic cosmological model

China did not have astronomers in the real sense until the Ming and Qing Dynasties. Wang Xichan and Mei Wending are the most famous ones, who are proficient in both Chinese and Western astronomical learning. Wang considered himself to be an adherent official of the former Ming Dynasty. After the fall of the Ming, he was determined to retreat from the official-dom and led the life of a hermit like Gu Yanwu. Although Mei Wending did not assume an official post in the Manchu regime, he was a friend of Emperor Kangxi. Emperor Kangxi thought highly of Mei's astronomical expertise and conferred him an inscribed board "Ji Xue Can Wei". The emperor even gave him a book compiled in his name and asked for his comments. Though having different life experiences, the astronomical calendars of the two tianxue masters are highly valued by later genera-tions. Wang and Mei's research on Tycho's Geocentric Cosmology can be regarded as representative work of Chinese astronomers.

In his book *Wu Xing Xing Du Jie*, Wang Xichan puts forward his cosmological model as follows:

> *The deferent (bentian) of the five plants is within the big sphere with the sun at the center. But the five planets are at the circumference of the deferent (bentian) while the sun is at the center, skewed slightly upward. The sun moves together with the bentian and forms the solar orbit.*

Wang was not satisfied with the Tycho's Geocentric Cosmology being used as the theoretical basis of the *Chongzhen Almanac*, so he intended to replace it with the above-said model. However, Wang's "bentian" has actually been replaced by another concept. In the "*Chongzhen Almanac*" and other works discussing Western astronomy at that time, "bentian" is a common term, referring to the circumference a celestial body runs on, corresponding to the "deferent" in the Ptolemy system. But Wang's "bentian" is in the eccentric position. In specific astronomical projections, bentian has virtually no effect; what really matters is the solar orbit, which happens to be the solar orbit in the Tycho model. Therefore, Wang's model does not differ from Tycho's. When Qian Xizuo commented on Wang's model, he pointed out that "although it appears different from Westerners', they are not mutually exclusive".[22]

Why did Wang Xichan deliberately behave to give the impression that he was different from Westerners? It had something to do with the political and ideological background back then.[23] Wang was an adherent of the former Ming Dynasty and refused to take an official post after the fall of the Ming. He showed strong discontent to Manchu's takeover of China, the regime's promulgation of Western astronomy, and the appointment of Western missionaries to take charge of the Qin Tian Jian (The Imperial Board of Astronomy). Compared with the traditional Chinese astronomical methods, the Western astronomy introduced at that time had obvious superiority in accurately estimating the celestial phenomena, which Wang found hard to accept. He firmly believed that the underlying cause for the traditional Chinese astronomical approach to fall behind was that there were no master astronomers in the court that could make best use of Chinese traditional methods. To this end, he produced *Xiao An Xin Fa* (*Xiao An New Calendar*), the last classical calendar in the Chinese history. In this book, he tried to incorporate some specific methods of Western astronomy under the premise of retaining the traditional Chinese calendar structure. However, his attempt was far from producing the desired effect, and the work became a particularly difficult book to read.[24]

The cosmic model in Mei Wending's mind is in essence the same as the Ptolemy model. But Mei does not totally agree with Ptolemy (see below) on whether or not a celestial body has a material orbit. Mei disagreed with the most important principle of Tycho's heliocentric theory. In *Mei Wu'an Xiansheng Li Suan Quan Shu-Wu Xing Ji Yao*,[25] Mei argues that "the five planets and the bentian (deferent) take the earth as the center". But in order not to run counter to the king's Tycho Model, Mei made a comprise and proposed the so-called "circumferential image of the sun", which holds that

---

[22] Qian Xizuo. Postscript to *Wu Xing Xing Du Jie.*

[23] See Jiang Xiaoyuan. "Wang Xichan's Life, Thoughts and Astronomical Activities", *Journal of Dialectics of Nature*, Vol. 11, Issue 4 (1989).

[24] See Jiang Xiaoyuan. "Wang Xichan and His *Xiao An Xin Fa*", *"Chinese Science and Technology Historical Materials"*, Vol. 9 Issue 6 (1986).

[25] Jianjitang block-printed version.

"the circumferential image" is a reflection in the human eye of the actual condition of the universe in Ptolemy's and Tycho's models.

If the trajectory of the planets on the "suilun" (epicycle) is connected, it will form a circle that takes the sun as the center. This is similar to the belief of new Western theories that the five planets center around the earth. The "circumferential image of the sun" is formed by the circular moving of the epicycle, the center of the epicycle runs on the circumference of bentian (deferent) and bentian centers around the earth. The three are interdependent. As they are not two, we cannot say they are different. Due to insufficient observations, some astronomers think that the five planets really center around the sun, which is a blunder.

Mei's *suilun* is equivalent to the "epicycle" in the Ptolemy model. At the beginning, Mei Wending applied the "circumferential image of the sun" to the outer planets. Later, his disciple Liu Yungong proposed that the theory also be applied to the inner planets, which was much lauded by Mei.[26]

As far as the consistency of the system is concerned, Mei's compromise is somewhat ingenious. He himself believes that his theory is in line with the gist of Tycho's model. "Once I made a drawing to calculate and found what underlies Tycho's theory is that it takes the earth as the center, which his theory explicitly states". Later, there came Jiang Yong who held in great esteem Mei's theory. In Volume 6 of his *Mathematics*, Jiang Yong used geometric methods to demonstrate Mei's model. He got identical results when calculating the celestial longitude in Mei's model by placing the planets in the suilun (epicycle) or the "circumferential image of the sun". The same is true with internal and outer planets.

As a matter of fact, Jiang Yong failed to prove the equivalence property of the Mei's model and the Tycho model in the *Chongzhen Almanac*. And Mei himself was not able to provide observation data to verify his model (Mei barely conducted any astronomical observations). In fact,

---

[26] Mei Wending commented. "I learned that my student Liu Yungong discovered that Venus and Mercury have epicycles, which is an unprecedented discovery. The evidence is evident and cannot be changed." See *Mei Wu'an Xiansheng Li Suan Quan Shu-Wu Xing Ji Yao*.

although Mei's model is ingenious, it is far from what the Tycho model intends to convey.

9. The spread of Copernicus' heliocentric system in China

As mentioned in the previous chapter, although the *Chongzhen Almanac* does not directly introduce Copernicus's model of the universe, it still touches upon some important contents of Copernicus's heliocentric theory. It was generally believed that Copernicus' theory did not enter China until the Jesuit P. Michel Benoist presented Emperor Qianlong the *Kun Yu Quan Tu* (*Atlas Maior*) in 1760.

The explanatory notes in *Atlas Maior* presented by P. Michel Benoist were later polished by Qian Daxin, and the atlas was published with a new name *Di Qiu Tu Shuo* (*Illustrated Geography*). It has a preface written by Ruan Yuan, in which not one word was mentioned about Copernicus's heliocentric theory, but it was repeatedly stated that the spherical-earth theory was credible and that the theory had long existed in China since ancient times. In the end, he said:

> The translated *Di Qiu Tu Shuo* crows about foreign customs and geography, which might be assumptions only. It talks about the system of the sun, the moon and the five planets, which is simply borrowed from Chinese tianxue philosophy. The underlying cause is that Chinese tianxue scholars scatter around the world. ... The theory is an old one Zhou Gong, Shang Gao, Confucius and Zeng Cen believe in. Our scholars should not take it as something new or dismiss it simply because it is "new".

Atlas Maior was not intended to explain the cosmological model. It seems that there is nothing wrong with Ruan Yuan's focus on the spherical earth. It is obvious that the Copernican universe model is incompatible with the official Tycho model. However, Ruan Yuan tried his best to dodge this issue. The arbitrary remarks that Copernican doctrine is borrowed from the old theory put forward by Chinese tianxue scholars are simply truism as the result of the prevailing *"Xi Xue Zhong Yuan"* (Western learning originated from China) that originated in the Qing Dynasty (see below). Obviously, such remarks are made out of a misinterpretation of Copernicus's theory. As a matter of fact, Ruan Yuan opposed the Copernican theory. Since he prefaced the *"Kun Yu Tu Shuo"*, he could not go so far as to attack the book directly. In *Jiang You Ren Zhuan Lun (Michael Benoist's Theory)* of *Chou Ren Zhuan* (*Vol. 46*), Ruan Yuan bluntly refuted the Copernican model and labeled it as a heretical doctrine that is "dislocated, inverted, indecent and worthless".

Ruan Yuan enjoyed a long life. When he compiled the *"Chou Ren Zhuan"* in 1799, he explicitly rejected the Copernican theory. But more than forty years later, he seemed to have changed his opinion and started to eulogize the heliocentric theory. In the preface of *"Sequel to Chou Ren Zhuan"*, he says:

> *I often think of Zhang Heng's seismograph, which has not been passed down. It was said that it could forecast earthquakes. But actually, it could not. I think Zhang Heng's device is based on the assumption that the earth moves while the sky is static. This could be the source of Michael Benoist earth-moving theory, or it might be coincidence.*

It is unreasonable to speculate that Zhang Heng's Houfeng seismograph in the Han Dynasty was the instrument to demonstrate the Copernican model. However, it was true that Western instruments of this nature were already introduced to the Qing court.[27] So, it was possible that Ruan Yuan got some inspirations from this fact. On the other hand, since the introduction of Western astronomical mathematics into China in the late Ming Dynasty, *"Xi Xue Zhong Yuan"* theory emerged, which was favored by Emperor Kangxi and his courtiers. That was why Ruan Yuan's assumptions had some lingering influence. Anyway, in such social context, it is not surprising that Ruan Yuan put forward the above-mentioned groundless speculation.

Compared with the Qing scholars' research on the universe models of Tycho and Ptolemy, the Qing people's discussion of the Copernican model always remained superficial. It was probably because this model was

---

[27] See Xi Zezong *et al.* "The Heliocentric Theory in China", *Science China*, Vol. 16, No. 3 (1973).

introduced to the Qing court rather late, and the Qing scholars' research on astronomy ebbed. On the other hand, celestial mechanics was at its prime, and the Copernican model had completed its historical mission.

10. Arguments over the truth of the cosmological models

Since the *"Chongzhen Almanac"* introduced the Western cosmic model and the small-circle (deferent) system, there appeared the question about the authenticity of the model and system. The *Chongzhen Almanac* holds an evasive attitude towards this issue. See the *Wu Wei Li Zhi* for further details (1).

All astronomical theories, no matter the celestial movement theory or small-circle (deferent) theory and concentric circle theory, are based on the brightness of the celestial bodies, not on their actual movement. That's why there are so many different theories. It's necessary to probe further to see which is correct and which is wrong.

This leaves more room for Chinese scholars to debate on this issue. The above question can actually be put in two ways: Broadly, does this type of cosmological model reflect the real situation in the universe? Narrowly, are the small circles, eccentric circles, etc., celestial bodies? Obviously, those affirmative to the narrow-sense questions must also make a positive answer to the broad ones, which can be called "authentic celestial body school"; those making negative answers to the narrow-definition questions can still have a different answer to the broad-definition ones, which can be categorized into "authentic non-body school" and "sheer assumption school".

In the Qing Dynasty, there were not many astronomers of the "authentic celestial body school", except Wang Xichan and Mei Wending. In his *Wu Xing Xing Du Jie (On the Movement of the Five Stars)*, Wang Xichan states explicitly:

If the five stars/planets move in the sky, they are real celestial bodies.

Wang Xichan failed to mention explicitly whether it was a 3D sphere or a 2D circular ring when he was talking about bentian. But in the eyes of Mei Wending and other Qing astronomers, bentian refers to a 2D ring. Mei argues that fujianlun wheel and suilun are virtual and intangible, while bentian is tangible hard circle (*Mei Wu'an Xian Sheng Li Suan Quan Shu-Wu Xing Ji Yao*).

There were others who can be categorized into "authentic non-body school", such as Jiang Yong, for in his *Mathematics* (Vol. 6), he says:

> *Though without a form when moving in the sky, it has the spirit of a circle. So, it can be seen as an authentic celestial body.*

This view holds that the Western cosmological model (mainly Tycho Model) reflects the real picture of the universe, only stating that the small circles and the eccentric circles are not authentic bodies.

What is most worth noting is the "sheer assumption school", which is the common viewpoint shared by most eminent scholars of Confucianism in the Qianlong and Jiaqing periods. For instance, Jiao Xun says in his *Jiao Shi Cong Shu-Shi Lun*:

> *The wheels in the sky are identified through actual observations, which does not mean the wheels really exist.*

This viewpoint is also shared by Ruan Yuan, who says in Volume 46 Jiang You Ren Zhuan Lun (Michael Benoist's Theory) of *Chou Ren Zhuan* that:

> *The circles are all assumption images based on addition and subtraction of the mean. But ignorant people… may believe that they are real objects in the sky, which is totally wrong.*

The person who states it most explicitly is Qian Daxin. He says:

> *Both epicycles and deferents are illusions, …. so are ellipses. If for the purpose of calculating to see whether the course of the stars and eclipses match the observations, it is feasible to name them big circles and small circles and even ellipses. (Qian Daxin, Chou Ren Zhuan-Vol. 49)*

Since all the circles are illusions, what is the truth? It is unknown and unknowable. The "sheer assumption" theory has similarities with

Ptolemy's "geometric representation", but it is not exactly the same. Ptolemy, Copernicus, Tycho, and others believe that their cosmological models reflect the real picture in the large structure and the specific small circles and the like may not necessarily exist; for example, Ptolemy categorized epicycles and eccentric circles into "circumference hypothesis mode".[28] This is between the "authentic non-celestial body" theory and "sheer assumption" theory. Ever since Kepler and Newton it has become the exact "authentic non-celestial body" school, whereas the "sheer assumption" theory is more rooted in the traditional Chinese concept of astronomy. The traditional Chinese method is to use algebraic methods to describe the movement of celestial bodies, paying little attention to the actual orbiting of celestial bodies.

Common sense tells us that those who are most interested in the operation mechanism of the cosmic model should be those in favor of the "authentic celestial body" school, which is the fact. The *Chongzhen Almanac* briefly introduces the idea of a magnetic attraction between celestial bodies, which has attracted the attention of some Chinese astronomers. This magnetic gravitational thought was once mistakenly thought to be from Chinese scholars or Tycho, but actually it is from Kepler.[29] Those who did further research and reasoning are mainly Wang Xichan and Mei Wending.

In *Wu Xing Xing Du Jie (Expounding the Movement of Five Planets)*, Wang Xichan tried to explain the circular motions of the sun, the moon, and the five major planets with the magnetic attraction between the celestial bodies.

> *When the seven celestial bodies (the sun, the moon and the five planets) move to the apogee and the perigee, they are harnessed by the primum mobile, which is the mediating master. Its energy controls the seven celestial bodies, like magnet to needles. When a planet moves to a certain place, it rises when it goes towards it and descends when it moves back from it.*

---

[28] Ptolemy: *Almagest*, 1X2, Great Books of the Western World, *Encyclopedia Britannica*, Vol. 16, (1980), p. 270.

[29] For details, please see Jiang Xiaoyuan. "Kepler's Celestial Gravity Thought in China", *Studies on the History of Natural Sciences*, Vol. 6 Issue 2 (1987).

He shifted the source of the magnetic pull from the sun as said in the *Chongzhen Almanac* to primum mobile, which might be attributed to the influence of the Aristotle model.

Mei Wending made more reflections on using magnetic gravitation to explain planetary motion. He used it to support his "circumferential image of the sun" theory.

> *Tycho says the sun attracts the five planets in the way a magnet attracts metal. So the distances from the planets to the sun are certain. Hence, when the sun completes a revolution, the trajectories of the planets around it take oval shapes. This is the principle with the new map of Tycho. Those unaware of this believe that the planets center around the sun, which is totally wrong (Mei Wu'an Xian Sheng Li Suan Quan Shu-Wu Xing Ji Yao).*

It was due to a misunderstanding that Mei Wending attributed the magnetic pull theory to Tycho. His explanation above does not agree with the cosmological model in his mind. Since the five planets are governed by the sun and center around the sun, the five stars already have physical links with the sun. Then how can the "circumferential image of the sun" be seen as virtual.

In general, the assumptions of Wang and Mei are still in a naive stage, far from scientific doctrines. In addition, there are still many others who hold on to the magnetic gravity theory in the Qing Dynasty. However, they only quoted words from the *"Chongzhen Almanac"*, with an academic level below that of Wang and Mei.

Chapter 12

Collision between the East and the West during the Ming and Qing Dynasties

1. The Xi Yuan Zhong Yuan (Western-learning-originated-from-China) theory put forward by Ming adherents

That the Jesuits kept introducing Western astronomy, mathematics, and other sciences and technologies to China fascinated a great number of eminent scholars, such as Xu Guangqi, Li Zhizao, and Yang Tingyun. After the Qing regime was founded, the Jesuit-compiled "*Chongzhen Almanac*" was renamed "Western New Almanac" and promulgated. Jesuit missionaries were appointed to preside over The Imperial Board of Astronomy for a long time. Emperor Kangxi himself even learned Western astronomy, mathematics, and other knowledge from the Jesuits. All of these exerted a great impact on the traditional beliefs and thoughts of Chinese scholars. "*Xi Xue Zhong Yuan*", prevalent in Chinese imperial courts and intelligentsia, was a product of that impact.

The idea "*Xi Xue Zhong Yuan*" was mainly about astronomical calendars. It also involves mathematics as it is closely related to astronomy. By the late Qing Dynasty, it had been extended to almost all areas of knowledge. However, it has obviously lost the value for historical scientific research, and so should not be included in the discussion here.

"*Xi Xue Zhong Yuan*" was conjectured by Ming adherents. According to available historical documents, the earliest advocate of the "*Xi Xue Zhong Yuan*" could be taken as Huang Zongxi. Huang accomplished a lot in Chinese and Western astronomical calendar studies and authored a great number of works on the astronomical calendar, such as Shou Shi Li Fa Jia Ru, and Xi Yang Li Fa Jia Ru. After the collapse of the Ming, Huang raised an army to fight against the Manchu troops. After he was defeated, he was exiled to the southeast coast. Even in such a difficult and precarious situation, he gave lectures in boats about annotated calendars. Huang once remarked: "It was said that Gougu (Pythagorean triples) was passed down by Shang Gao of the Zhou Dynasty, but Westerners just stole it." He was referring to mathematics, but scholars then regarded mathematics and calendar as one. Huang's concept of *Xi Xue Zhong Yuan* was affirmed by Quan Zuwang in Volume 11 of his *Jie Qi Ting Ji* (*Jieqiting Collection*). He noted in the article "*Li Zhou Xian Sheng Shen Dao Bei Wen*" [*Divine Sacred Inscriptions of Mr. Li Zhou (Huang Zongxi)*]:

Later, Mei Wending discussed the origin of Western calendars based on *Zhou Bi Suan Jing* as calendar studies, which amazed the world. People thought it was his creative viewpoint. But they didn't know it was proposed by Huang Zongxi in the first place.

Another pioneer advocating "*Xi Xue Zhong Yuan*" was Fang Yizhi, a contemporary of Huang Zongxi. As a *jinshi* (a successful candidate in the highest imperial examinations) in the 13th year of the Chongzhen Period, Fang was exiled to the south after the collapse of the Ming Dynasty. He had been loyal to the Yongli Regime and threw himself into the anti-Manchu struggle. His work "*Fu Shan Wen Ji*" (*Fushan Collection*) was banned in the early Qing Dynasty, and therefore few copies were in circulation. In the article You Zi Liu Xu, after his discussion of Chinese ancient astronomical calendars by Fang Yizhi, said (though many of his remarks are not from layman perspectives):

During the Wanli Period [of the Ming Dynasty], Chinese people were much enlightened due to the coming of Western scholars. Their maps are so well illustrated that you get to know the laws the moment you open them. People may amaze at them, but they do not know that our sages had said it. ... Confucius once remarked:" [*Due to political turbulence*] *the Emperor lost his officials taking charge of various fields of knowledge, making the knowledge scatter in remote areas.*[1]

Fang's argument that "the Emperor lost his officials taking charge of various fields of knowledge, making the knowledge scatter in remote areas" is worth noting as it coincides with the argument "the Court lost the rituals and ethics and had to seek for them among the folks" (to be elaborated later) held by Mei Wending and Ruan Yuan in later periods.

Huang and Fang failed to provide any specific evidences to support the argument. The theory did not make much progress until Wang Xichan (1628–1682) expounded it further. When the Ming Dynasty was collapsing, Wang made two attempts at suicide. After he was saved by others, he refused to collaborate with the new regime and regarded himself as an adherent of the former regime. He was an important member of the Adherent Circle, with Gu Yanwu as the leader. Wang dived into studies of astrological and calendar studies. Later generations saw him and Mei Wending as two top astronomers of the Qing Dynasty. Wang was adept at both traditional astronomy of China and Western astronomy that made its way to China at the end of the Ming Dynasty. His achievement in subjects of astronomy is believed to be much higher than that of Huang Zongxi and even higher than that of Fang Yizhi. He repeatedly discussed *Xi Xue Zhong Yuan*. The most important part is as follows:

> *Few people believe that Western astronomical calendar dwarfs Chinese tianxue and Chinese students of astronomy find it astounding. Some scholars like the grotesque nature of Western astronomy and flaunt it, believing China is short of it. But as a matter of fact, all are stated in Chinese tianxue. There is nothing special with Western astronomy.*
>
> *One is to use pingqi and dingqi to gauge zhongjie (mid point), which is similar to what is said in Chinese Tianxue, such as shourenshi, sizheng methods.*
>
> *One is to use the apogee and the perigee to measure the movement of the moon, which is similar to the descriptions of lunar eclipses in Chinese tianxue.*

---

[1] Fang Yizhi: Fu Shan Wen Ji Hou Bian in *Qing Shi Zi Liao* (*Studies in Qing History*) (*Vol. 6*), Zhonghua Book Company, (Vol. 2) (1985).

> One is to use zhenhui and shihui to measure eclipses, which is nothing different from what Chinese tianxue says about eclipses.
>
> One is use xiaolun and suilun to measure the five stars, which is similar to the methods of using pinghe, dinghe, chenxi, fujian, jichi and liutui of Chinese tianxue.
>
> One is to use nanbei didu (south and north latitude degrees) to measure the north pole and the dongxi didu (east and west longitude degrees) to mark time zones, which is similar to that in Chinese tianxue.
>
> Probably when our forefathers determined a tianxue law, they followed its own rationale. The law is elaborate while the rationale is not, as the rationale is concealed in the law. Industrious and prudent scholars can always reason for the rationale. Western astronomers stole our ideas, but they could never go beyond.[2]

These remarks prove to be the most important for the development of the *Xi Xue Zhong Yuan* argument. These words were uttered not long before 1663, close to the time Huang and Fang held the view. For the first time, Wang Xichan provided specific evidences for a theory, which proved to be false. The theory involved the movement of the sun, the moon, the planets, solar and lunar eclipses, fixing of solar terms, and time service, including almost all major aspects of traditional Chinese calendars. Wang acknowledged that the Western calendar was better than traditional Chinese calendar in many aspects, but said that it still used the inherited Chinese methods of calendar compiling, which was nothing new.

It stands to reason since he argued that Western methods of calendar compiling had long been in use in China since ancient times, there was a possibility that the two had been developed independently by the Chinese and the Westerners and that and there were coincidences between the two. But Wang Xichan flatly ruled out this possibility, believing that Westerners stole Chinese methods secretly. This assumption paved the way for Mei Wending to come up with his own theory.

It is worth noting that Huang, Fang, and Wang, who were loyal adherents of the Ming and were influential figures in history, happened to stick

---

[2] Wang Xichan. *Li Ce,* published in *Chou Ren Zhuan* (Vol. 35).

to the same argument of *Xi Xue Zhong Yuan*. This should not be deemed accidental in the history of science.

2. Emperor Kangxi's call

After the Qing regime was established, Emperor Kangxi was, on one hand, obsessed with the Western science and technology introduced by the Jesuits, and, on the other, advocated the *Xi Xue Zhong Yuan* theory. Emperor Kangxi wrote a book named imperial San Jiao Xing Lun (Triangle Theory), in which he said "Chinese ancient calendar spread to the West and Western people practiced it and made it even better". This remark is about the astronomical calendar. In mathematics, his *Xi Xue Zhong Yuan* theory is even more eye-catching. A much-quoted line was about Emperor Kangxi and Zhao Hongxie talking about algebra in 1711 was rcorded by *Dong Hua Lu*:

> *Western arithmetic is perfect, which was originally Chinese. They called*
> *it Algebra, which means "coming from China".*

It is generally believed that the name Algebra, written in Chinese characters in different ways, is transliterated from the Arabic word Al-jabr, meaning "algebra". It remains unclear from which of these facts Emperor Kangxi drew the conclusion that Western algebra (Dong Lai Fa) came from China. Some people think it might be because the word Al-jabr is confused with another Arabic word Aerhjepala in pronunciation. It is also questioned whether Emperor Kangxi had come into contact with Arabic and whether he had learned some Arabic from the Jesuits serving in the inner court — the same Jesuits who had taught him Western astronomy and mathematics (as normally they used Manchu language and Chinese language). To say the least, even if Algebra means Dong Lai Fa ("methods coming from the east"), it is still not a convincing enough answer to the question regarding when the Chinese methods were introduced to the West. Later, Mei Wen Ding assumed the responsibility of answering this question.

According to the documents about the Jesuits working in China, Emperor Kangxi started to learn Western astronomical mathematics in 1689 from the Jesuits. Since then, he was obsessed with Western sciences so much so that he attended four hours of lectures per day for four consecutive years. After class, he even did exercises.[3] In the following decades, Emperor Kangxi enjoyed flaunting his knowledge about astronomical mathematics and geography. Probably, the *Xi Xue Zhong Yuan* theory was proposed by the emperor himself as he had had some knowledge about Western astronomical mathematics.

So far we have not found any documents saying that Emperor Kangxi had ever studied the books of Huang, Fang, and Wang. As the three scholars refused to work with the Qing regime, Emperor Kangxi was not likely to read their works, as the emperor had so many matters to attend to.

After all, Emperor Kangxi did not have much attainment in astronomical calendar; but he did understand some of the Western astronomical mathematics, not much however. This can be seen from his talks on astronomical calendar with his courtiers recorded in his book Ji Xia Ge Wu Bian. Emperor Kangxi believed that he himself could evaluate Mei Wending's *Li Xue Yi Wen* (*Doubts about Astronomical Calendar Studies*). As Mei Wending, though not assuming any post in the new regime, was willing to cooperate with the Qing regime in academic studies, Emperor Kangxi could willingly spare some time to read his book. But the book was not a profound book. In contrast, Huang Zongxi and Wang Xichan achieved much higher attainments in Western astronomical studies. If it was they who proposed the idea of *Xi Xue Zhong Yuan*, it could be that they saw the similarities between Chinese and Western astronomy. If it was Emperor Kangxi who put forward the idea, he must have done that for political reasons.

3. Mei Wending's keen response to Emperor Kangxi's call

Emperor Kangxi's advocacy met with keen response from Mei Wending, who repeatedly exalted the theory.

The imperial *San Jiao Xing Lun* (*On Triangles*) states that Western learning originated from China. How great is his majesty! His wisdom

---

[3] De Fontaneg. "Letters on February 17, 1703", "*History of the Qing Dynasty*" (6th edition), Zhonghua Book Company, (1985), pp. 161–162.

outshines that of so many scholars. [Yu Zuo Shan Chuang, Volume 4 of *Ji Xue Tang Shi Chao (Poetry and Prose Anthology of Jixue Hall)*].

Respectfully, I've read the imperial *San Jiao Xing Lun* which says ancient Chinese calendar spread to the West and Western people practiced it and improved it. How smart is the Emperor saying so, which can pacify controversies on this issue. [Shang Xiao Gan Xiang Guo 3/4, Ji Xue Tang Shi Chao (*Poetry and Prose Anthology of Jixue Hall*)].

I have respectfully read *San Jiao Xing Lun*. It says curves can be calculated with angles and the lengths to the central point. This method has long existed in China, only that it was lost after it was introduced to the West. Western people learned it and elaborated on it. How smart is the Emperor saying so. It can serve as the golden rule in compiling calendars. (*Li Xue Yi Wen Bu*, Vol. 1).

Respectfully and humbly, Mei Wending read the imperial *San Jiao Xing Lun*. He not only immediately handed over the copyright of the "*Xi Xue Zhong Yuan*" to the emperor (See the words "How great is his majesty! His wisdom outshines that of so many scholars". The theory had been proposed by Huang, Fang and Wang much earlier. It was impossible for Emperor not to know the three scholars. It was also beyond imagination if Mei Wending did not know the three people.) but set up his mind to replenish and improve "*Xi Xue Zhong Yuan*" theory. In Volume 1 of *Li Xue Yi Wen Bu*, he expounded the issue from the following three aspects.

First, the argument "Hun Tian Tong Xian" was formerly the theory of Gaitian ("canopy-heavens") in *Zhou Bi Suan Jing*. At the end of the Ming Dynasty, Li Zhizhu wrote the book *Hun Gai Tong Xian Tu Shuo,* and the Jesuit Sabatino de Ursis wrote *Jian Ping Yi Shuo*. The former talks about the projection in the plane of the spherical coordinate network and introduces the astrolabe and its use; the latter introduces the astroscope called Jianping, which is similar to astrolabe. Mei Wending based his argument on the Hun Tian Tong Xian.

So huntian (celestial sphere of upside position) is like a sculpture while gaitian (celestial sphere upside down) is like a painting. .. If we know that gaitian and huntian are not two different things, we know the Western calendar and ancient Chinese calendar are of the same origin.

The theory of canopy heavens illustrates the three-dimensional universe with plane techniques. Though the device is plane, its representations are three-dimensional. Hence, the Hun Gai Tong Xian theory undoubtedly evolved from the ancient Gatian Theory.

Research shows that the five cold and hot zones in Western calendar are in conformity with the seven belts (Qiheng) in Zhou Bi Suan Jing. Though the latter does not mention the earth is round, the calculations are based on the assumption that the earth is round. Hence, the star map in Western calendar is also inherited from the theory of canopy heavens.

After discussing such evidences, the five zones, spherical earth, and star map, Mei Wending asserted:

> *As for the devices to gauge the huntian and gaitian spheres, ... they must have been made by sages like Rong Cheng and Li Shou. As what the Western calendar talks about corresponds with what Zhou Bi Suan Jing states, we can safely assert that it belongs to the gaitian device stated in Zhou Bi Suan Jing.*
>
> *As the Jianping device uses plane circles to gauge the celestial sphere, it can be regarded as a Gaitian device.*

It is not difficult to see that Mei's starting point of this argument is totally wrong. The ancient Huntian theory (theory of sphere-heavens) and Gaitian theory (theory of canopy heavens) in China are by no means compared to sculpture and painting. Li Zhizao learned the astrolabe principles from the Jesuits and wrote the *Hun Gai Tong Xian Tu Shuo*. He only borrowed the ancient Chinese terms of Huntian and Gaitian, with the actual content being completely different. Proficient in astronomy as Mei was, it is impossible for him not to understand this. He deliberately gave strained interpretations and drew far-fetched analogies only to draw favor from the Emperor, a typical feudal scholar's practice, which can never be convincing. As for sages like Rong Cheng and Li Shou, whether they existed in history remains a big problem, let alone the question of whether they produced a huntian and gaitian spherical device capable of projecting spherical coordinates to a plane. Therefore, the proofs of five zones, spherical earth, and star map are all far-fetched analogies.

Second, let us take a look at the channels by which Chinese calendar compiling methods were introduced to China. *Xi Xue Zhong Yuan* can only justify itself with a clear idea of how Chinese calendar making methods were introduced to the West. We can start from the record of the

Shih Chi-Li Shu (Shih Chi-Calendar), which says "after King You's and King Li's reign, the Zhou Dynasty was on the decline. ... So the astronomers went away, to various states in the Central Plains and remote areas", Mei Wending believed that "to run away from social upheaval, many scholars and astronomers fled to other states with books, documents and apparatuses". However, Mei Wending envisioned another channel that was more advanced. *"Shangshu-Yao Dian"* says that "Emperor Yao ordered Xihe to watch closely the movement of the sun and time and Xizhong, Xishu, Hezhong and Heshu to station in four quarters of the territory to observe the celestial phenomena and make the astronomical calendar". Based on this folklore, Mei envisaged that there were oceans on the southeast and extreme cold in the north that were obstacles for the spreading of Chinese calendar expertise, and only the west direction was barrier-free. "It can go west and further westward". In this way, the Gaitian theory of *Zhou Bi Suan Jing* spread to the West via the Western route. He further envisaged that when Hezhong was heading westward, the knowledge of Emperor Yao and Emperor Shun spread to the countries around. When Hezhong arrived in the Western world, he saw the following spectacular scene:

> *Folks came from remote areas to see him. They admired Hezhong for his virtue and talent. Some learned from his teachings and passed down what they learned from him. Some got inspirations from him and got enlightened. Some brilliant ones developed their own theories based on what they had learned from him.*

According to Mei Wending's speculations, the profound, grand, and well-structured Western astronomy was developed from the oral teachings of Hezhong in a few words. Of course, compared to Wang Xichan's assertion that Western calendar methods were stolen from China, Mei's oral teachings do not sound that fancy.

Third, let us take a look at the kinship between the Western calendar and Hui Hui Calendar, namely Islamic astronomical calendar. Mei Wending uttered the following words:

> *Westerners, adept at calculating, developed the Hui Hui Calendar and made it more precise. Therefore, Western calendar and Hui Hui*

Calendar are not much different from each other, as both of them origi-
nated from the Theory of Canopy Heavens. This is the result of Chinese
calendar spreading to the West. Some Westerners got the full picture of
Chinese calendar while others didn't. Some studied it closely while oth-
ers only got a rough idea of it. But the root was the same.

Mei Wending was able to see the kinship between Western astronomy
and Islamic astronomy, which was much more difficult than we do today,
as Chinese scholars back then knew very little of the outside world. But,
Mei Wending was wrong in the precedence relationship. It was true that
the Western calendar was more developed than the Hui Hui Calendar. But
if we trace back, we will find that the latter originated from the former.
Among the three evidences Mei used in his demonstration of the *Xi Xue
Zhong Yuan* theory, only the third one is tinted with scientificity, though it
has nothing to do with the subject he tried to prove.

With Emperor Kangxi's advocacy and Mei Wending's arduous elabo-
ration, the *"Xi Xue Zhong Yuan"* appeared more persuasive, and its influence spread wider.

4. Many scholars add "fuel to the fire"

With the wholehearted advocacy of the great Emperor Kangxi and the
promotion of Mei Wending (who was famed as the No. 1 scholar in calendar studies of the Qing Dynasty) who had written in various genres, the
"Xi Xue Zhong Yuan" argument spread near and far, and nobody dared to
challenge it. Volume 1 named *Zhou Bi Jing Jie* (*Interpretations of Zhou Bi
Suan Jing*) of the masterpiece *Shu Li Jing Yun* (*Essential Principles of*
Mathematics) made with the imperial order in 1721 has the following
words:

Johann Adam Schall von Bell, Ferdinand Verbiest, Antoine Thomas and
Domingo Fernández Navarrete had successively engaged themselves in
astronomical calendar studies and elaborated mathematics as well, as a
result, the principles for illustrating numbers of degrees were getting
more detailed. However, if asked where they are from, they would say
they originated from China.

But it remains doubtful whether the above-mentioned people had really uttered such words, on what occasions they had said so, and in what kind of context they had said so. At least, the occasions and the context remain doubtful. If what *Shu Li Jing Yun* says were true, that would be precious material for the study of the relationship between Emperor Kangxi and the Jesuits. The Jesuits who served in the Qing court received much courtesy, but after all they were courtiers of Emperor Kangxi. When the emperor had said so, they had no way out but to agree with him.

Ming Shi (*The History of the Ming Dynasty*) was completed in 1739. Its *Li Zhi* (*Calendars*) reiterated the assumptions of Mei Wending and Hezhong's journey to the West. It went on to elaborate this as follows:

> *They extensively collected sporadic learning and expertise scattered among the folks for thousands of years, as they were only interested in the low-end knowledge in the populace.*

This kind of self-complacence was very popular among Chinese scholars. When the Qian-Jia School was at its prime, its representative figure Ruan Yuan exerted great efforts to exalt the *Xi Xue Zhong Yuan* theory. In *Chou Ren Zhuan* (*Biographies of Astronomers*) compiled in 1799, Ruan Yuan repeatedly discussed *Xi Xue Zhong Yuan* theory. There are some innovations in his discussions. For instance, he said in Volume 45 of Biography of Johann Adam Schall von Bell the following remarks:

> *However, I (Ruan Yuan) have ever read history, chronicals, astronomy and mathematics extensively and studied various schools of astronomy, astrology and arithmetics. I have got to know that the new calendar is a work encompassing all the merits of different schools of the past and the present, rather than a total novelty conjured by the Western compilers. For instance, Zeng Cen (505–435B.C.), also called Zeng Zi, states that the earth is round in his ten articles; the apogee and the perigee theory (Tai Yang Gao Bei) coincides with what is said in Kao Ling Yao (the so-called Di You Si You argument); the different solar terms theory coincides with Jiang Ji's theory on calendaricity; the theory of stars and planets revolving at orbits of different heights is nothing different from Xi Meng's theory of celestial bodies revolving at different altitudes (Bu*

Fu Tian Ti). To sum up, how can they prove that the above theories did not originate from China? For instance, didn't Jiegenfang (algebra) come from China?

This saying is far-fetched, and inferior to Mei Wending's. The influence of the Qianjia (Qianlong-Jiaqing) School on the academic circles of the Qing Dynasty was well known. With the preaching of Ruan Yuan, *Xi Xue Zhong Yuan* theory has exerted a lasting effect. For example, in 1882 when the Qing Dynasty was coming to the end and the "*Xi Xue Zhong Yuan*" theory had been in existence for more than two centuries, Zha Jiting made the following remarks in the postscript of the reprint of *Chou Ren Zhuan*.

All those amazed at expertise of Western astronomers got to know that "the round Earth theory" comes from Zeng Cen after they read Ruan's and Luo's books. The "nine-layer heaven" comes from Tian Wen (Asking the Heaven), the "triangle and octalines" comes from Zhou Bi Suan Jing; the different solar terms theory originates from Jiang Ji from the late Qin Dynasty; the waxing and waning of the moon originates from Zu Chongzhi of the Qi Dynasty; the "Huntian and Gaitian Being One" theory comes from Cui Ling'en of the Liang Dynasty; the Jiuzhi technique comes from the translation of Gautama Siddha in the Tang Dynasty; the Jiegengfa (algebra) comes from Qin Jiushao in the Song Dynasty and the Tianyuan (equation) technique from Li Zhi in the Yuan Dynasty. The Western method is insignificant. But when we trace back to the sources, we find they all originated from China.

When the "*Xi Xue Zhong Yuan*" theory had been established, it found its way into astronomy, mathematics, and other disciplines. For example, Ruan Yuan said in the article about the chime clock in Volume 3 of *Yan Jing Shi San Ji* that the Western chime clock was no different from the ancient Chinese clepsydra, and therefore it originated from China. This is the influence of this theory in the field of mechanical engineering. In medical science, Mao Xianglin said that Western surgery forms an organic whole with Hua Tuo's techniques of treating injuries and fractures. He went on to say as Westerners failed to get the quintessence, that the success rate was not high [*Mo Yu Lu (Vol. 7)*]. Most of these remarks are assumptions of laymen, and therefore are of no academic value at all.

5. Social background of Xi Yuan Zhong Yuan

Adherents of the Ming Dynasty and the Emperor and his courtiers of the Qing Dynasty were antagonistic politically, but the two groups of people invariably advocated the "*Xi Xue Zhong Yuan*" theory, which is quite a noteworthy phenomenon.

The astronomical strife in the history between China and the West began in late Ming Dynasty. Prior to this, China had been exposed to Western astronomy twice, in the Six Dynasties, Sui and Tang dynasties, and the Yuan and Ming dynasties, respectively. But the two introductions were done indirectly. In the first, Indian astronomy was the medium; the second time around, Islamic astronomy was the medium. Moreover, Chinese astronomy was then quite developed and there was no possibility for it to be replaced by astronomies from other lands. So, there was no dispute between China and the West. In the Yuan Dynasty, there were concurrently two observatories, Huihui and Han'er, and in the Ming Dynasty the Imperial Board of Astronomy had a special Huihui section, which used the *Hui Hui Calendar* and the *Da Tong Li* (*Datong Calendar*) as references for each other. At that time, there was no strife between the two calendars either.

But when the Jesuits came to China in the late Ming Dynasty, Western astronomy had developed to a very high level. Comparatively speaking, Chinese traditional astronomy lagged far behind. The Ming Court decided to compile the "*Chongzhen Almanac*", which was thought to replace the millennial traditional calendar with a Western calendar. As a calendar had long been seen as the symbol of ruling power of a dynasty since ancient times, this became intolerable for many Confucian scholars who had long taken pride in the heavenly dynasty. Did it mean that such a sacred calendar would be made with Western methods? Did it mean that a Western calendar would be turned into a Chinese one? It was because of this that the *Chongzhen Almanac*, since the compiling work was started, had been under attack from the conservatives. Owing to the efforts of Xu Guangqi, the *Chongzhen Almanac* was finally completed in 1634. But the attack by the conservatives impeded the spread of the almanac even 10 years after its completion.

Right after its founding, the Qing regime promulgated the revised *Chongzhen Almanac* in the name of *Xi Yang Xin Fa Li Shu* (*Western New*

Almanac). This time, the courtiers and scholars did not have different opinions towards the adoption of Western calendar, and this was because of two facts. On the one hand, following the dynasty change, the new regime in China would always change the calendar to show a new world is born and a new dynasty has a new mandate of heaven. At that time, there was no other calendar than *Chongzhen Almanac* that was better than the *Da Tong Li* (*Datong Calendar*). On the other hand, as it was not long since the Manchu people occupied the Central Plains of China as alien forces, by no means could they have claimed themselves to be descendants of Emperor Yan (Yandi) and the Yellow Emperor (Huangdi). They were actually foreigners to the land, not much different from Westerners. So, they just used whatever was useful to them. As Joseph Needham noted later: "After the regime change, Johann Adam Schall von Bell felt he was allowed to use West and Western freely in the court, as Manchu people came from outside".[4]

Since the Manchu took the throne in China, there was no longer any taboo for being barbarian. In 1729, Emperor Yongzheng pretended to be fully at ease in saying: "My dynasty doesn't have any taboos for any barbarian names in *Da Yi Jiao Mi Lu* (*Record of an Awoke Scholar, Vol. 1*)". He only cited Mencius' words: Emperor Shun came from the barbarian ethnic group in the east while King Wen of the Zhou came from the barbarian ethnic group in the west", stressing only the virtuous one could become an emperor and it did not matter whether he came from, be it the barbarian ethnic groups or the Central Plains.[5] But as a matter of fact since they occupied the Central Plains, the Manchus fully accepted the Han culture. Besides, the new regime had lasted for two generations, and the Han scholars' pains of being subjugated were fading away. At this time, unconsciously the Manchu rulers had regarded themselves as legitimate rulers of China. This change was the ideological background for Emperor Kangxi to advocate *Xi Xue Zhong Yuan*.

On the other hand, the idea "*Xi Xue Zhong Yuan*" was put forward by Huang, Fang, and Wang, who were nurtured by time-honored traditional

---

[4] See *Astronomy, Science and Civilization in China,* Vol. 4, p. 674.
[5] See Emperor Yongzheng's remarks in Vol. 1 of *Da Yi Jiao Mi Lu* (*Record of an Awoke Scholar*), in the fourth edition of *Studies in the Qing History*, Zhonghua Book Company, (1983).

Chinese culture and were most loyal to the former Ming Dynasty. They witnessed Manchus' occupation of the Central Plains and their full acceptance of Western astronomers and Western techniques in calendar making, something they thought most sacred, causing dual discontent in their hearts. Among them, Wang Xichan was the most representative figure. He had lived under the rule of the Qing regime for decades, and in the depths of his mind he had always wanted to see the Chinese traditional calendar adopted by the new regime. Of course, some Western astronomical calendar expertise could be borrowed to make up the weaknesses of Chinese calendar. "We can feed in Western materials in the mould of Datong Calendar", he said. For this purpose, he made tremendous efforts to pick out the weaknesses of Western calendar and meanwhile demonstrated the rationality of *"Xi Xue Zhong Yuan"*. Finally, he concluded:

> *All those places in the new calendar that go against those in the old calendar are not so good and all those that are a bit better than those in the old calendar all originated from the old calendar. [Li Ce-Calendar].*

His six-volume *Xiao An Xin Fa* is an incarnation of his proposition. As the Ming-adherent scholars were determined not to cooperate with the Manchu regime, they were unable to make suggestions to the Manchu regime on calendar issues. So, they had to resort to *Xi Xue Zhong Yuan* to get them out of the theoretical predicament. The influence of traditional culture made them adhere to the ideal of "turning Chinese into Western", but the harsh reality was "turning Western into Chinese". If they could prove "the Western originated from China", they would be able to evade the conflicts between the two.

The Yang Guangxian Incident in the first year of the Kangxi Period revealed the seriousness of the "Western–Chinese" calendar issue. This incident can be seen as the repercussion of the Western–Chinese strife on astronomy. Yang's being sentenced marked the failure of his last efforts in proving Chinese calendar excelling Western calendar. Yang uttered a famous line in his book *Bu De Yi*, which goes as follows:

> *I would rather China not have a good calendar than having Westerners serving in Chinese court.*

It is clear that he did not put the merits of the calendar itself in the first place. He attempted to attack their calendar since the Jesuits took the astronomical calendar as the vehicle to get promoted. After tasting crushing defeat many times in contests of actual observations with Ferdinand Verbiest, Yang resorted to the ideology to attack him.

> The calendar I stick to is passed down by Emperor Yao and Emperor Shun. The throne of the Emperor was inherited from Emperor Yao and Emperor Shun. The tradition His Majesty inherits is from Emperor Yao and Emperor Shun. Therefore, the calendar His Majesty issues for enforcement must be the calendar passed down by Emperor Yao and Emperor Shun.[6]

Though Yang was eventually found guilty and dismissed from office, he received a lot of sympathy from orthodox scholars. They were sympathetic with him mainly beause of his support for traditional Chinese culture. Therefore, the issue became a theoretical dilemma that the court must tackle.

The dilemma confronting the rulers of the Qing dynasty was that, on one hand, they badly needed Western learning, they needed Western astronomy to make a calendar, needed Jesuits to help with diplomacy (such as the signing of the Sino-Russian Treaty of Nerchinsk), needed Western technology to produce artillery and other instruments, and needed quinine to cure the diseases of the royals, among other things; on the other hand, they were proud of being the inheritors of the millennia of traditional culture, conceited to stand for the "Heavenly Empire". Therefore, in something as sacred as the calendar symbolizing the imperial power, adopting a Western calendar became a pressing, nagging issue for the Emperor and his courtiers. Under such circumstances, Emperor Kangxi advocated the "Xi Xue Zhong Yuan" theory, which is after all a clever way to get out of that dilemma. In this way, the Qing rulers successfully made use of Western scientific and technological achievements, and meanwhile freed themselves from the suspicion of "turning Western methods into Chinese". Though the Western calendar was superior, it originated from China after all. Only they developed it better. Therefore, adopting Western calendar was a practice of "regaining what was lost from the commoners".

---

[6]Huang Bolu. Zheng Jiao Feng Bao (Imperial Praise of the Orthodox Religion), the Tz'umu t'ang of Shanghai, (1904), p. 48.

Another reason for the popularity of *"Xi Xue Zhong Yuan"* among Chinese scholars for 300 years is that the advocates expected to take it as a vehicle to improve the national self-esteem and enhance national self-confidence. For thousands of years, China used to be the "Heavenly Country", spreading knowledge to other lands and having nations come to pay tributes. They felt embarrassed that all of sudden they found themselves inferior to others in many things. Ruan Yuan's remarks are typical of this:

> *If our Chinese calendar is dwarfed by Western calendar, then should we consult Europeans on calendar issues since our great Qing Dynasty has inherited a lot from ancient calendars? Absolutely not! Those adept at algorithm must be fully aware of that. (Biography of Johann Adam Schall von Bell, Chou Ren Zhuan Vol. 45).*

But, the reality was ruthless indeed. The calendar in use during the 260 years of the Qing regime was truly from the West. Though the *"Xi Xue Zhong Yuan"* theory intoxicated many scholars for the time being, with the lapse of time this illusion would soon clear away.

6. Xu Guangqi and Fang Yizhi

In the history of ideas of the Ming and Qing Dynasties, Xu Guangqi should be counted as one of the most important figures. Due to the social division of labor back at that time, Xu Guangqi did not appear as a scientist in modern sense, but in fact he was a fully qualified astronomer, mathematician, and agronomist.

From the perspective of the history of science, Xu Guangqi was one of the rare people of foresight in that era. Because he had earnestly studied Western sciences that began to take modern forms, he had his own judgment of the merits and demerits of Chinese and Western academics. He even uttered some remarks to belittle the traditional Chinese astronomical mathematics, such as:

> *Following the questions and answers by Shang Gao, Rong Fang asked Chen Zi about the dimensions of the sun, the moon, the heaven and the earth, which are big follies. [Gou Gu Yi Xu (On Pythagorean) in Vol. 2, "Xu Guang Qi Ji (Anthology of Xu Guangqi)].*

> *In the chapter Pythagorean of "Jiu Zhang Suan Shu (The Nine Chapters on the Mathematical Art) there are indeed several illustrations using tables and squares for measurement, which are somewhat similar to those in Ce Liang Fa Yi (Principles of Measurement). Though the methods used are somewhat similar, the meanings are totally erroneous. Scholars can hardly discern them. [Ce Liang Yi Tong Xu Yan (Preface on Similarities and Differences of Measurement) in Vol. 2, "Xu Guang Qi Ji (Anthology of Xu Guangqi)].*

These remarks were later under fierce attack by supporters of *Xi Xue Zhong Yuan* in the Qing Dynasty. The source of *"Xi Xue Zhong Yuan"*, the Theory of Canopy Heavens of the *Zhou Bi Suan Jing,* was dismissed a "big folly" by Xu, which enraged scholars like Mei Wending and Ruan Yuan. After a comparison of *Jiu Zhang Suan Shu (The Nine Chapters on the Mathematical Art)* and *Ji He Yuan Ben (Euclid's Elements)* Mei Wending also found the superiority and the inclusiveness of *"Jiu Zhang Suan Shu"* [*Wu An Li Suan Shu Mu-Yong Gou Gu Jie Ji He Yuan Ben Zhi Gen (Wuan's Calendar Booklist-Using Pythagorean to Find the Root of Euclid's Elements)*].

This is because Xu Guangqi was on a completely different ideological realm from Mei Wending and his counterparts. In Xu Guangqi's mind, there was no strife between the Chinese and the Western, and he only enthusiastically called for the advent of new sciences and made unremitting efforts to spread these new sciences. Xu was once called "Chinese Bacon". Poetic as it may sound, it basically makes sense. As for Mei Wending and his colleagues, as seen above, their academic activities were tinted with political color to a large extent. Emperor Kangxi set the task for them, that is, help the court find the way out of dilemma of "Turning Western into Chinese" and "Turning Chinese into Western". They made up their mind to do everything possible to win their self-esteem (face) under the circumstances that China had lagged far behind the West in science and technology. Such a sentiment of scholars like Mei Wending has influenced modern scholars, who believed that Xu Guangqi's derogatory remarks about Chinese traditional astronomical mathematics had gone too far. The problem is that they do not give a thought to the historical background and significance of Xu's airing of his views.

With regard to Xu Guangqi's attitude towards Western learning, there is a complicated legal case worthy of noting. In the process of presiding over the making of "*Chongzhen Almanac*", Xu Guangqi presented a series of memorials to the emperor. In the *Li Shu Zong Mu Biao* (*Table of Contents of Calendar*), he said:

> If there is any inaccurate understanding during translation, I will have people check the corresponding contents in the Da Tong Li and ask those who have a good understanding of Da Tong Li so as to get an accurate understanding.
>
> I would integrate Western materials into the mould of Da Tong Li. It is like building a house. I will use the same dimensions and specifications and high-quality columns, stone materials, tiles and bricks. The house thus built will stand for thousands of years without any possibilities of collapsing.

This passage sounds very much like the earliest version of "Chinese learning should be followed as the essence; Western learning as the practical application". Xu said that the *Chongzhen Almanac* would be fully in accordance with the pattern of traditional Chinese calendar, with only some parts (tiles and bricks) taken from the Western astronomy. However, when it was completed, it was completely Western, no matter in theory or in application. This provided his critics with a handle. Wang Xichan was a representative of his critics, and his heckling in *Xiao An Xin Fa-Zi Xu* (*Preface in Xiaoan' New Calendar*) goes as follows:

> When he first began his translating, he said he would "integrate Western materials into the mould of Da Tong Li". But it came out that he discarded the original calendar and adopted the Western calendar.

Xu Guangqi's remarks can only be seen as an improvised makeshift tactic as he encountered intense resistance when he worked on the new calendar. He said this to appease the conservatives so that his work on the new calendar could go on smoothly. That was why his acts belied his words.

While trying to completely promote new sciences, Xu Guangqi was so foresighted enough as to realize the dross of the traditional Chinese culture that had been entwined with science. Fang Hao noticed that when a lunar eclipse occurred Xu Guangqi pleaded to the Emperor, saying that he was not able to attend the "Rescue" ritual as he had observations to attend to. The "Rescue" was a kind of ancient ritual in China that had lasted for thousands of years. The purpose was to pray to the Providence not to let the sun and the moon get hurt during eclipses and to forgive the faults of the Son of Heaven (the Emperor) on earth. Fang Hao held that Xu Guangqi was making an excuse to evade the ritual.

It must be due to the fact that Xu Guangqi had become a believer of Catholicism that Xu was reluctant to attend the Rescue during the lunar eclipse. According to Catholic canon, believers must not attend superstitious activities. So he chose not to attend it. But if it was true he didn't attend it because he had to observe the sky, nobody should be dubious of that.[7]

Whether Xu Guangqi did not attend the Rescue ritual because of commandments of the Catholic canon is open to discussion. At least he thought that scientific observation was more important than superstition. Among his contemporaries, Xu's zeal towards new sciences could only be matched by that of Wang Zheng and Li Zhizao. About half a century ago, Shao Lizi made the following comments about Xu and Wang.

Academy knows no national boundaries. We should learn the merits of others to make up our own demerits. We should catch up with the tides of new sciences in the world. Their love for the country, nation and truth was as pure as snow and as fervent as fire.[8]

Seen from modern perspectives, this comment can be still regarded most appropriate. Compared with Xu Guangqi, Fang Yizhi seems to have been higher in some modern academic works, which is still largely biased. This is because Fang once criticized Western learning whereas Xu Guangqi fervently praised Western learning. For a very long period of

---

[7] Fang Hao. *The History of Sino-foreign Relations*, Yuelu Publishing House, (1987), p. 705.

[8] Shao Lizi. "To Commemorate the 300th Anniversary of Wang Zheng's Death, *"Journal of Truth"*, Vol. 1, Issue 3 (1944).

time, our criteria to judge a person was whether he/she had ever criticized the West. If yes, he/she was good; if not, he/she was not good. Of course, this emotional bias still appears as a fair judgment. For instance, Fang is often praised for his sensible attitude towards Western learning: not total or uncritical acceptance, nor total rejection.

Fang's criticism of Western learning can be seen from the much quoted Zi Xu of *Wu Li Xiao Shi* (*Preface of Physics*), which states:

> *During the Wanli Period, Western scholars were adept at zhice (sciences), but not at tongji (philosophy), but the wise estimated that their actual observations were not consummate.*

This paragraph seems to have endorsed Western learning indeed. But as *tongji* (philosophy) is too vague a concept, what superiority does it have when compared with Western modern sciences highlighting analysis and experimentation? As recent researchers have pointed out:

> Even if zhice is interpreted as "science", it can hardly raise the judgment of Fang Yizhi. The spread of Western learning at the end of the Ming Dynasty was indeed tinted with medieval religious superstition. Besides, as modern sciences were at its booming formation period and knowledge was updated fast, it was natural for it to be not consummate. But who does "the Wise" refer to? if it refers to himself, we are not able to see from what scientific stance he pointed out that the Western learning was not yet consummate.[9]

For example, Fang's criticism of Matteo Ricci's sun–earth distance in his Li Lei of *Wu Li Xiao Shi* (*Calendars of Physics*) was cited as evidence that Fang criticized the Western learning and that Chinese learning was superior to the Western learning. As a matter of fact, his critique of Western learning was out of his misunderstanding of Matteo Ricci's *Qian Kun Ti Yi* (*Explication of the Structure of Heaven and Earth*).[10]

Fang's attitude toward Western learning was not much different from that of most traditional Chinese literati at that time. Chinese culture's

---

[9] Fan Hongye. *"The Jesuits and the Chinese Science"*, Renmin University of China Press, (1992), p. 141.

[10] Fan Hongye. *"The Jesuits and the Chinese Science"*, (1992), pp. 141–142.

supreme chauvinism had been entrenched in their mind. It was not coincidence that Fang became one of the pioneers of *Xi Xue Zhong Yuan* theory. It is worth noting that the illusion of Chinese tongji (philosophy) being superior to Western zhice (sciences) is still deeply rooted in the hearts of many Chinese people.

7. Criticism and dispute on *Xi Xue Zhong Yuan*

Though the "*Xi Xue Zhong Yuan*" theory was rampant during the Qing Dynasty, there was still no short of skeptical critics. Jiang Yong was one of them.

Jiang Yong was a master in Confucian classics of the Qing Dynasty and had superior knowledge in astronomy and mathematics as well. He wrote a work *Mathematics* elaborating classical Western astronomical system. The work, consisting of six volumes, is also named *Yi Mei*, which is said to be presented to Mei Wending. At that time, Mei Juecheng, the grandson of Mei Wending, alias Xunzhai, was in the Emperor Kangxi's good graces. He was an advocate of the *Xi Xue Zhong Yuan* theory, too. After reading Jiang Yong's work, he composed a poem for Jiang.

Though you have great achievement in studying European astronomy, we should not forget the words of Mencius.

He expressed his views in the name of Mencius, which Jiang Yong fully understood. He said in the Preface of his Mathematics:

> *By the poem, Mr. Xunzhai meant to warn me not to forget the contributions of our forefathers since the Western astronomy, as the latecomer, has surpassed Chinese astronomy. He hoped I would not advocate Western calendar too much and I should use traditional Chinese astronomical methods passed down by Xihe more often.*

Here, Jiang Yong was talking about Western astronomy that had surpassed Chinese astronomy as a latecomer and the traditional methods adopted by Goddess Xihe in ancient times. As for this advice, Jiang Yong flatly refuted it. He said:

> *Today, the Western astronomy has been prospering like the sun at high noon. Who have opened up the trail to modern astronomy? Western astronomers. Their efforts in blazing the trail shall never be forgotten.*

This small paragraph of Jiang Yong is concise and comprehensive, and, in fact, systematically refuted the "*Xi Xue Zhong Yuan*" theory: First, Jiang Yong denied the proposition that Western astronomy originated from China, but stressed Western astronomers' merits in establishing modern astronomy. Second, Jiang Yong flatly rejected Mei's proposition of attributing the achievements of Western astronomers to ancient Chinese tianxue people. Third, he acknowledged that Western astronomers were able to establish astronomy that is better than Chinese. This means he disapproved the belief that Chinese culture is higher than other cultures, on which the "*Xi Xue Zhong Yuan*" argument is based.

Soon after, many more eminent scholars joined in the debate. As recorded in Volume 49 of *Chou Ren Zhuan* (*Biographies of Astronomers*), Jiang Yong's student Dai Zhen praised Jiang, saying "Jiang's attainment in astronomical studies is not lower than that of Mei Wending". Qian Daxin, after reading Jiang Yong's "*Mathematics*", wrote a long letter to Dai Zhen despising Jiang. He said: "I have long heard that Mei Wending does not like Jiang Yong's theory and I suspected that he deliberately suppressed Jiang. Today I've read Jiang's book and got to know that Mei Wending has well inherited the family education tradition and his knowledge well dwarfs Jiang Yong's". He even scolded Dai Zhen asking him if he spoke highly of Jiang just because he was his student. The book *Mathematics* is of course not free from errors, but Qian Daxin's discontent was mainly triggered by Jiang's refusal to join in the "*Xi Xue Zhong Yuan*" chorus.

With an open mind, Jiang Yong was not solitary amid eminent scholars at that time. For example, Zhao Yi said in Vol. 2 of his "*Yan Pu Za Ji*" (*Yanpu Miscellanies*)" that Western astronomy is better than Chinese tianxue and that it was created by Westerners themselves.

Most of the documents and books for divinations and calendar making used in the Imperial Board of Astronomy are written by Westerners as their calculations are much more accurate than traditional Chinese methods. The Xuanji and Qizheng theory (the theory of the sun, the moon and the five planets) appearing sacred, started to tell the secrets of the heaven and the earth. The West, though thousands of miles away, has better astronomical expertise. They know how vast the universe is. There are many sages skilled in astronomy around the world, in addition to Xihe, Xuanyuan, Chao and Suiren in ancient China.

Zhao Yi was also an open-minded scholar. He took a pragmatic view on Western learning.

8. Merits and demerits of Emperor Kangxi

In recent years, there have emerged some works on historiography highly judging Emperor Kangxi. In rare gifts and bold strategy, they compare him to the French "Sun King" Louis XIV; in support of academy, they portray him as an outstanding figure comparing him to Florence's Cosimo Medici (Cosimo Medici) during the Renaissance. Seen from the letters and reports to Europe by the Jesuits who served in the court of Emperor Kangxi, he was often praised as "benevolent", "impartial", "generous", "wise", and "great". It was true that Emperor Kangxi was interested in Western science and technology and was also eager to learn Western science and technology. It was rare for the traditional Chinese feudal society to have such a monarch. As an individual, he can be indeed counted as a Chinese person with foresight in vision and knowledge during that era. However, as the head of a big state, he did have historical merits and demerits. First, let us look at Emperor Kangxi's recruiting of Jesuits expert in astronomy to work in the court, as it is often seen by scholars as important evidence that the Emperor loved science or was enthusiastic about science and technology. But if we see it in a broader context in the long history of ancient China, Emperor Kangxi was not much different from his predecessors and successors. Chinese dynasties had the custom of recruiting people adept in occult techniques to serve in the court, and they were mostly monks or Taoist priests. They attended upon the emperor with their occult techniques, such as divination by astrology, auguring, medicine, alchemy, painting and calligraphy, music, and so on. In general, they were like hangers-on. However, when they had won the trust of the Emperor, it was highly likely that they would attend to affairs of national defense and administration. The Jesuits serving as courtiers in Emperor Kangxi's administration conforms exactly to this model. Though the Jesuits were no astrologists or alchemists, they served the courts with medicine, painting, music, and other crafts. In addition, they provided Western instruments to maintain chime clocks and designed Western-style royal architectures. The expertise and matters they were engaged in might be different, but the pattern was no different from previous dynasties. Emperors recruited Western people to work in the court not because they loved science but because they took it as honor in having them around.

Emperor Kangxi had a more serious fault as pointed out by previous scholars, that is, he was interested in Western technology himself, but did not have any intention to make it known to his courtiers and the common folks.

It seemed that Emperor Kangxi wanted to make use of or enjoy Western learning only, as he didn't make efforts to nurture talents of astronomy or change the academic climate. Liang Qichao once criticized him by saying: "Even if he did not deliberately stifle wisdom of people, it can be counted as his fault". As for me, I think Emperor Kangxi certainly deserves such criticism.[11]

Emperor Kangxi even never tried to have children of the Eight Banners learn technology from the Jesuits serving in the court — which was something not difficult to do at all. He also did not establish public schools for the Jesuits to impart Western technology or send young scholars to Europe to study. These things were not difficult to do at all, and were undoubtedly what the Jesuits were very happy to do.

When modern sciences were budding, Emperor Kangxi had a once-in-a-lifetime opportunity to enable China to keep pace with Europe in technological development. Emperor Kangxi was on the throne for 60 years. By rights, he had every condition to promote Western learning in China. But, unfortunately, his thinking, on the whole, was still restrained in the old pattern. His so-called "open mind" only stayed at the superficial level. He was only content with learning something rare that was unknown to ordinary people.

Emperor Kangxi failed to see the flush of dawn of a new world.

---

[11] Shao Lizi. "To Commemorate the 300th Anniversary of Wang Zheng's Death, *"Journal of Truth"*, Vol. 1, Issue 2, (1944).

9. Was there any scientific revolution in the 17th century?

A few years ago, N. Sivin put forward a moving view in an article with many versions that the scientific revolution in China began in the 17th century; he said:

> *(In the 17th century) Chinese astronomers began to believe that mathematical models could be used to explain and predict various astronomical phenomena. These changes are equal to a conceptual revolution in astronomy. ... (This revolution) is no less than Copernicus's conservative revolution, but cannot be compared with Galileo's radical mathematicalized hypothesis.* [12]

But in fact this argument is probably out of misunderstanding. It faces at least two problems. First of all, in terms of mathematical model, regardless of whether the traditional Chinese algebra method is a mathematical model or not, even after the introduction of the Western geometric model, many Chinese astronomers only thought of this model as a means of computing. According to Vol. 49 of *Chou Ren Zhuan*, Qian Daxin made the most concise remarks, which was most representative, and this states:

> *Both epicycle and deferent are false phenomena, which have been discarded. Today, we have created ellipticity, which is false, too. If we are to check the distances of solar and lunar movement and whether solar and lunar eclipses are in conformity with estimations, we can use big or small cycles or ovals.*

They did not think that the geometric model of the West had any substantial significance. Ancient Chinese scholars did not show much interest in the authenticity of the universe and its operational mechanisms.

Second, a more serious problem is that since the widely accepted "*Xi Xue Zhong Yuan*" theory had asserted that Western astronomy

---

[12] N. Sivin. "Why did not China have a scientific revolution — or did it really happen?", "*Science and Philosophy*", Issue 1, (1984).

originated from China long ago, there is no need for substitution or change of concepts.

In the 17th-century Chinese scientific community, the most popular concept could be considered to be huitong (comprehend). In *Li Shu Zong Mu Biao*, Xu Guangqi proposed that "if we want to overtake, we must be able to huitong" (comprehend). We do not know what Xu means by "overtake", but we do know that he held the view that Chinese scholars must be able to huitong. He cherished the hope that through the research into both Chinese and Western astronomy, Chinese astronomy could catch up with and even overtake the West.

Later, Wang Xichan and Mei Wending were generally believed to be masters of both Chinese and Western astronomy. However, in the context of *Xi Xue Zhong Yuan* being promoted across the country, they went astray. They turned to prove the argument of *Xi Xue Zhong Yuan* theory, rather than gain a thorough understanding of Western astronomy. As Yabuuchi Kiyoshi pointed out:

> As the representative calendarist of the Qing Dynasty, Mei Wending mainly aimed to make a compromise between Chinese and Western astronomy. He did not assimilate Western astronomy comprehensively, nor did he intend to develop it.[13]

This is what China was like in the 17th century. Even if there were buds of scientific revolution, they were drowned by the tide of "*Xi Xue Zhong Yuan*".

---

[13]Yabuuchi Kiyoshi. "History of Science and Technology in the Ming and Qing Dynasties", *Science and Philosophy*, No. 1, (1984).

Chapter 13

Legacies of Chinese *Tianxue*

1. Most developed subjects of ancient China

Many books have concluded that there are four developed subjects in ancient China, namely, agriculture, medicine, astronomy (*tianxue*) and arithmetic. Although this might be true, from the perspective of scientific development, these four subjects have different weightages.

Ancient Chinese agriculture and medicine have not yet lost their vitality. Ancient agricultural theories and technologies are still valuable for today's agricultural production. The vitality of Traditional Chinese Medicine (TCM) is obvious to all. Western medicine has not yet become an exact science, so it is not a science yet.[1] Of course, the situation with TCM is no better, if not worse. However, TCM works pretty well for some diseases that Western medicine cannot handle. Therefore, TCM and Western medicine are complementary like horse and horse.

Astronomy (*tianxue*) and arithmetic in ancient China were deemed akin and closely related to each other. Things are similar in the West. Royal astronomers or astrologists in ancient times were formally addressed as mathematicians in the West. Today, Chinese mathematicians and astronomers accounted for scholars, although the official title is often "mathematician".

---

[1] In the Western discipline classification, science, mathematics and medical science are considered parallel, which means that the latter two are not part of science. This perspective is totally different from the normal perspective of the public in China.

Similar to their peers elsewhere in the world, Chinese mathematicians and astronomers are classified based on the Western system. Normally, we call them *modern mathematicians* or *modern astronomers*. Today, students of mathematics in China no longer read *Zhou Bi Suan Jing* (*The Arithmetic Classic of the Gnomon and the Circular Paths of Heaven*) or *Jiu Zhang Suan Shu* (*The Nine Chapters on the Mathematical Art*) and students of astronomy no longer read *Shih Chi-Tian Guan Shu* (*Records of the Grand Historian-Tianxue Officials*) or *Han Shu-Lu Li Zhi* (*History of the Han Dynasty-Calendar*). In this context, should you ask whether ancient mathematics and *tianxue* in ancient China had vitality or not, we'll have to say that the answer is definitely *No*. Today, we will talk about the legacies of ancient *tianxue* in China based on the above understanding.

2. Categories of the legacies of Chinese *tianxue*

It is not easy to answer the question: What on earth is the legacy of ancient Chinese *tianxue* (astronomy)? Chinese people often refer to them as "rich and precious legacy" left by Chinese ancestors. But what is the legacy and what purpose does it serve? How should we look at them? These questions are worth evaluating and considering, as few discussions have ever been found on these important questions.

First, with regard to the legacies of ancient Chinese *tianxue*, what pops up in our mind has already been mentioned in Section 2 of Chapter 5. The details include mostly the records of celestial phenomena in the book, *Sylloge of Records of Celestial Phenomena in Different Dynasties*, which has got over 10,000 records that are the most scientific part of the *tianxue* heritage. Although the ancients recorded these celestial phenomena for astrological purposes, they are still of reference value to modern astronomy. As modern astronomy takes celestial bodies as the object, the evolution of celestial bodies spans long periods of time and millions of years in merely a blink of the eye. In this sense, ancient records, even though not much scientific or accurate, are still precious.

Second, there are more than 90 kinds of calendars,[2] which are of the most scientific color in the *tianxue* legacies. The reason why astronomical

---

[2] Listed on pages 559–561 of *Sheet of Chinese Calendars* of *Encyclopedia of China* (*Astronomy*) are 93 kinds of calendars, 69 of which have documentary records.

records are of scientific value is that they can be used today; but as they were recorded for astrological purposes, they are so scientific. However, calendar is just the opposite. As the ancient Chinese calendar is for the study of the law of celestial movement and involves mostly mathematical astronomy, it reflects the knowledge of astronomy at that time, which is where the scientific color exists. However, it is color in sheer sense, as it in essence served astrology.[3] Since the vast majority of these calendars are useless today,[4] they are naturally not as scientific as astronomical records.

Third, it remains a question as to how to look at the large amount of *tianxue* literature discussed in Chapters 5 and 6 as well as the various fragmentary records in the ocean of Chinese ancient documents.

Perhaps, we can try to approach the legacies of Chinese *tianxue* from another perspective. We can classify these legacies into three categories:

Category 1: Legacies that can be used to solve issues of modern astronomy;
Category 2: Legacies that can be used to solve issues of historical chronology;
Category 3: Legacies that can be used to understand the legacies of ancient societies.

This classification can basically cover all legacies of Chinese *tianxue*. In the following sections, we will aim to reveal the features of these three categories through specific cases, which have been deeply obscured due to historical and professional estrangement.

3. Legacy one: Outbreaks of novae and supernovae

In the early 1940s, the Taurus' crab nebula was confirmed by astrophysicists as the remains of the supernova outburst in 1054. This outburst was

---

[3] Please refer to Chapter 4 of my book *Tianxue Zhenyuan* for a comprehensive demonstration.
[4] Today's lunar calendar was derived from *Shixian Calendar* of the Qing Dynasty, the last official calendar in Chinese history, whose theoretical foundation, however, was not based on Chinese *tianxue* but the European astronomy in the 16th and 17th centuries.

recorded in detail in ancient Chinese documents.[5] With the invention of radio telescope, an apparatus was used to observe the waveband beyond visible light, which was derived from the radio system popular in the Second World War. The Taurus' crab nebula was discovered as a very strong radio source in 1949. In the 1950s, radio sources were also discovered from the remains of the 1572 supernova (known as "Tycho Supernova", named after the famous European astronomer Tycho Brahe who mastered the meticulous observation of the supernova) and the 1604 supernova (also known as "Kepler Supernova"). Astronomers then came up with the following conjecture: supernova outbreak could form radio sources. Once a star bursts out, the brightness increases componentially to thousands of times instantly, hence, this phenomenon is called "nova outburst". If the outbreak is more intense, the brightness could increase by tens of millions or even hundreds of millions of times, known as the "supernova outburst". In this process of outburst, extremely massive material and energy are sprayed into the cosmic space. Humans on the earth could only spot a new bright star as the earth is so far away from the outburst. If the earth were closer, it would have been destroyed instantly. If that was the case, there would be no need for a study.

Fortunately, supernova outbursts are extremely rare. Take the solar system, for example, there have been 14 documented supernovae during a period of 2,000 years. Since 1604, there hasn't been any. Therefore, to verify the above assumptions of astronomers, we can only resort to historical records as it is impossible for us to wait for thousands of years. The former Soviet astronomers were much interested in this concept. For lack of historical data in the West, they had to turn to China for help. In 1953, the Soviet authorities of astronomy sent a letter to the Chinese Academy of Sciences (CAS), asking for information on supernova outbursts in history. At that time, CAS Vice President Zhu Kezhen delegated the mission to a young astronomer named Xi Zezong, who was later elected an academician.

Some foreign scholars once attempted to verify the records of supernova outbursts in history. An important figure was Lundmark, who published the *Suspected New Stars Recorded in Old Chronicles and among*

---

[5] For a comprehensive discussion, please refer to *The Historical Supernovae* by Clark and Stephenson or the Chinese version by Jiangsu Science and Technology Press, 1982.

Recent Meridian Observations in 1921. Until 1955, astronomers around the world had to resort to the catalogue when they referred to novae and supernovae in the ancient times. However, this catalogue had serious shortcomings in terms of accuracy and completeness.

From 1954 onwards, Xi Zezong published a number of papers, such as *Discussion of the Relationship between Supernova Outbursts and Radio Sources from Chinese Historical Records* and *The Relationship between Supernova Outbursts and Radio Sources in Chinese History*. Then, in 1955, he published *Gu Xin Xing Xin Biao* [*A New Catalogue of Ancient Novae*].[6] He made best use of the complete and accurate astrological records in ancient China, examined and corrected a total number of 90 nova and supernova outbursts from the Shang Dynasty to 1700.

Ten years later, in 1965, Xi Zezong and Bo Shuren published the sequel paper *Records of Novae in Ancient China, Korea and Japan and Their Significance in Radio Astronomy*.[7] In this paper, the co-authors amended *A New Catalogue of Ancient Novae* and based on supplementary historical documents from Korea and Japan worked out on a more complete chronological Catalogue of nova and supernova records, which still has a total number of 90 outbursts. This paper also proposed seven criteria for identifying star outbreaks from records of comets and other variable stars as well as two criteria for distinguishing supernova outbursts from nova outbursts. Meanwhile, it also discussed the frequency of supernova bursts based on the historical records.

A New Catalogue of Ancient Novae immediately caught the attention of the United States and the Soviet Union. Both countries reported it and translated the full text. It was easy to understand that the Soviet Union came up with such a response; but considering the strained relations between China and the West at that time, it is remarkable that the United States was so interested in this paper. Of course, American astronomers might not care about politics. Vice president Zhu Kezhen of CAS thought highly of the catalogue. He deemed this paper and the *China Seismic Data Chronology* as the two major achievements since studies of scientific history were initiated in the People's Republic of China. As a matter

---

[6] *Acta Astronomica Sinica*, Vol. 3, Issue 2, (1955).
[7] *Acta Astronomica Sinica*, Vol. 13, Issue 1, (1965).

of fact, with the development of astrophysics, *A New Catalogue of Ancient Novae* proved to be more important than what he had ever thought. In the following year, after the paper came out, there emerged two versions in English in the US. In the next two decades, the paper was cited for over 1,000 times by astronomers around the world when they discussed the latest development of supernovae, radio sources, pulsars, neutron stars, X-ray sources, and gamma-ray sources. *Sky & Telescope*, one of the most prestigious journals of astronomical studies in the world, commented thus:

"For Western scientists, perhaps the most famous articles on *Acta Astronomica Sinica* are the two articles on supernovae published by Xi Zezong in 1955 and 1965 respectively."[8]

The much-cited *Astronomy of the 20th Century* by Struve and Zebergs mentioned only one Chinese work, which is *A New Catalogue of Ancient Novae*.[9] It's worth lauding that a work is so frequently cited and is so much valued. Above all, it is so closely connected to a great many developments in the field. Why so? The answer can be found from the development of modern astronomy.

According to the modern stellar evolution theory, at the end of evolution, stars of different mass will form white dwarfs, neutron stars or black holes. How many stars will experience a nova or supernova outburst before evolving into a white dwarf star? One way to discuss this problem is to calculate the frequency of the supernova outbursts based on historical records. The theory of stellar evolution also predicts the existence of neutron stars composed of ultra-dense material. In 1967, Hewish discovered the pulsars, which were soon confirmed to be neutron stars, thus confirming the prediction of the stellar evolution theory. Many astronomers believe that neutron stars are the remains of supernova outbursts. Black holes, though not observable directly, can be indirectly identified. The Cygnus X-1 is an X-ray radiation source that is considered to be one of the celestial bodies most likely to become black holes, and some

---

[8] *Sky & Telescope*, Vol. 10, (1997).
[9] O. Struve and V. Zebergs, *Astronomy of the 20th Century* (Crowell, Collier and Macmillan, New York, 1962).

astronomers suggest that the celestial body corresponds to the historical records of supernova eruptions. Later, astronomers also found that after outburst, supernova will form an X-ray source and γ-ray source. These advances in astrophysics and high-energy physics are all related to the supernova eruption and its remains, and thus could not have been without the historical records of the supernova eruption. This is the underlying reason why *A New Catalogue of Ancient Novae* and its sequel article have long been valued by astronomers of various countries.

Having said so much about the Chart without citing equations, but including quite a lot of jargons, I certainly have not forgotten that we are talking about heritages. The above only shows how great the scientific value of the supernova outburst records in ancient China is. But what do the original historical records look like? It's better to take a look at them, right? Well, let's take a look at the records of the supernova occurring in 1054:

1. On July 4, 1054, a nova appeared a few inches to the southeast of Tianguan (Star ζ of Taurus) and disappeared more than one year later (*Song Shi-Tian Wen Zhi: History of the Song Dynasty — Astronomy*).
2. In the third month of the first year of Jiayou Period (1056), Sitianjian (imperial astronomer) said: "Since the fifth lunar month of the first year of Zhihe Period (1054), a nova appeared in the east, close to Tianguan (Star ζ of Taurus) and disappeared one year later" (Song Shi-Renzong Benji: *History of the Song Dynasty — Basic Annals of Emperor Renzong*).
3. In July 1054, a nova appeared a few inches to the southeast of Tianguan (Star ζ of Taurus) and disappeared in the third lunar month of the first year of the Jiayou Period (*Xu Zi Zhi Tong Jian Chang Bian Vol. 176: Sequel to 'Comprehensive Mirror in Aid of Governance'* (First Draft)).
4. On the 22nd of the seventh month of the first year of the Zhihe Period (1054), the commanding general Zuo Jian wrote to an official named Yang Weide thus: "I saw a guest star (nova) appear. It is tinted with yellowish color." *Huang Di Zhang Wo Zhan (Yellow Emperor's Divination)* stated thus: "Though the guest star is bright, it does not go against Bisu (Constellation Taurus). If it is bright, it means the host country is to have a sage. I plead to let it known to the national

archives so that all the officials can congratulate on it. So, an imperial edict came, asking to have the information delivered to the national archives" (*Song Hui Yao: Compendium of Government and Social Institutions of the Song Dynasty*, Vol. 52).

5. In the third month of the first year of the Jiayou Period (1056), the Sitianjian (imperial astronomer) said: "That the guest star is gone is the omen of a saga leaving. In the beginning, in the fifth lunar month of the first year of Zhihe Period (1054), it appeared in the east, close to Tianguan (Star ζ of Taurus). In the morning, it looked as bright as Venus in daytime, with white and reddish rays shooting out. The guest star lingered there for 23 days" (*Song Hui Yao: Compendium of Government and Social Institutions of the Song Dynasty*, Vol. 52).

These are the original historical records that possess extremely high scientific value. The fourth one is particularly interesting: a retired *tianxue* official Yang Weide, who had long held an important post in the imperial astronomical authority sent a memorial to the emperor, claiming that based on astrological theories, the supernova outburst meant that the country was soon to have a sage and pleading to send the relevant records to the national archives and let all the officials congratulate on it. The emperor did give the consent to it. In this way, scientific documents were concealed in the astrological literature.

4. Legacy two: The color of Sirius (Accelerators)

We must not take it for granted that the heritages left for modern astronomy only exist in the above-mentioned 10,000 records of celestial phenomena. In fact, they also exist in other literature. If you are not capable of extracting something of worth from a miscellany, you can never discover these heritages.

Sirius, or Star α of the Canis Major, is the brightest star in the celestial sphere, giving off dazzling whiteness. It is also a visual binary (according to astronomical tradition of the academia, the major star is called Star A while the companion star called Star B), and Star B is the earliest confirmed white dwarf. But such a famous star has been perplexing scholars of current stellar evolution theory just because of the description of its color in ancient records.

In the Western ancient literature, Sirius is often described as red. Scholars have identified similar descriptions in ancient Babylonian wedge clay tablets as well as works of Ptolemy, Seneca, Cicero, Flaccus in ancient Greek-Roman times. In 1985, Schlosser and Bergmann said they had found the work of Gregory, the bishop of Tours (now in France) produced in the sixth century in an early medieval manuscript, in which he mentioned a red star that can be identified as Sirius. So, they asserted that Sirius had been red until the end of the sixth century before it became white. Their article was published in the authoritative journal *Nature*,[10] sparking a new round of controversy and concern about the color of Sirius. As of 1990, *Nature* published at least six papers in response to the controversy.

In accordance with the current theory of stellar evolution and today's understanding of the two stars of Sirius, it was impossible for its A-star to change color over a period of 1000 or 2000 years. If Sirius was truly red in the 6th century, the only possibility was the B-star, as the B-star is a white dwarf, and stars go through the giant red star stage before evolving into white dwarfs. Hopefully, this could explain the records of Sirius turning red. It could be the grand red light of the B-star that overshadowed the A-star. However, according to the current theory of stellar evolution, the time required for a red giant star to turn into a white dwarf is longer than 1500 years, allowing for the extreme situations. Therefore, we can hardly find a ration explanation for the Western ancient record of Sirius turning red.

So, astronomers are left only with two alternatives: either questioning the current stellar evolution theory or denying the authenticity of the ancient record of red Sirius.

Indeed, the Western records of red Sirius is not impeccable in terms of authenticity. As Seneca was a philosopher, Cicero a political commentator and Horace a poet, their astronomical attainments are hard to verify.

---

[10] Schlosser, W. and Bergmann, W., *Nature*, Vol. 318, (1985), p. 45.

It is true that Ptolemy is a great astronomer and astrologer, however, there is still room for questioning in many details of his theory.

As for the red star described by Bishop Gregory, many people believe it was actually not Sirius but Dajiao (or Arcturus in the West, which is Star α of bootes).[11] Arcturus is just a bright red star.

Since the Western record is confusing, can ancient Chinese records of *tianxue,* which is so bountiful, provide forceful evidences to settle this case? I have thought of it for a long time, but as China boasts an ocean of historical documents, looking through such a huge volume would be impossible.

There was no astrophysics in ancient times and ancients would not observe the color of celestial bodies with today's light. Whenever Chinese ancient documents talked about stars and planets, without exception, they all focused on the astrological meaning of these colors. In most cases, these records do not have any scientific significance on addressing the issue of Sirius color. These records are also usually in the same format. Two examples are as follows:

> "To the east is a big star called Wolf (Sirius). If the angle-type rays change color, it means there will be many thieves and brigands. (Shih Chi-Tian Guan Shu)"

> "Wolf (Sirius).... If the rays and angles change color, there will be wars; if it is bright, weaponry will become too dear to buy. ...If its color turns lustrously yellow, something auspicious will happen; if the color is black, something lamentable will happen."

Obviously, Sirius can change color at any time, sometimes yellow and sometimes black, and even black according to some divinations. It even oscillates. Modern astronomical knowledge tells us that it is absolutely impossible. But ancient Chinese astrological literature abound in similar divinations about fixed stars, with things signaled different though. To address the issue of Sirius changing color in ancient times, there is no point resorting to records of this kind, which are even misleading. For

---

[11] For example: Mc Cluskey, S. C., *Nature*, Vol. 325, (1987), p. 87; van Gent, R. H., *Nature*, Vol. 325, (1987), p. 87, all believe what Gregory described is Dajiao (Arcturus).

instance, Gry and Bonnet-Bidaud committed such a mistake in the two articles published in *Nature*,[12] as they based their argument on the line, "if the angle-type rays change color, it means there will be many thieves and brigands", in *Shih Chi-Tian Guan Shu*, asserting that Sirius is changing color. They meant to use it to clear away the counterexample of Sirius in the current stellar evolution theory of Sirius; but they failed to note that there exist ancient divinations on color changes of other stars. If they were changing colors according to their deduction, hundreds of new counter examples would arise, which means the current stellar evolution theory of stars would be invalid.

Yet, providence doesn't let down a man who does his best. After four or five years of careful research, I finally found another record of Sirius changing color in the ancient Chinese astrological literature. These kinds of records, though limited in number, are the most reliable, which is really a matter for rejoice.

So, it turns out that ancient Chinese astrologers not only believed that stars often changing the color had an astrological significance but also assumed that the same divinations also apply to planets. To ascertain the different colors of planets, they set the standards for the colors; Specifically, they designated a few famous stars to be the standard stars for different colors. As a matter of fact, the opportunity to solve the problem of Sirius's color change is hidden here.

As for the color of Venus, in *Shih Chi-Tian Guan Shu*, Sima Qian set the five-color standards as follows: "Sirius is white, Antares is red, Betelgeuse (α Ori) is yellow, Bellatrix (γ Ori) is blue, Mirach is black."

The five stars above are as follows, respectively: Sirius, Antares (Scorpio α), Betelgeuse (Orion α), Bellatrix (Orion γ), Mirach (Andromeda β). Except for Sirius whose color needs to be ascertained, the ancient records of the colors of the remaining four stars are credible.

The red standard star Antares is ascertained as red today whereas the blue standard star Bellatrix is considered to be blue; the yellow standard star Betelgeuse is proved to be a red giant star, which, modern scholars believe, could be possibly yellow 2000 years ago according to modern

---

[12] Gry, C. and Bonnet-Bidaud, J. M., *Nature*, Vol. 347, (1990), p. 625.

stellar evolution theory.[13] The black standard star Mirach has been proven to be dark red. The ancients were definite about it as black is reasonable. China boasts the profound thought of the five elements, which have its influence in various aspects of life. The practice of classifying stars into five stars is an inevitable manifestation of the combination of the five elements thinking and astrology. The regular pattern of the five colors matching with the five elements is inevitably white (Venus in the west), red (Mars in the south), yellow (Saturn in the center), blue (Jupiter in the east), and black (Mercury in the north).

So, black color is a necessity. As the five-star standard was used for observations in nighttime, if a star is truly black, it would be impossible to be spotted and compared. Hence, as an alternative, dark red was used to mark it instead.

From the above study on the colors of the four stars, we can find that Sima Qian's account of the five-color standards is credible. So, from the four-star color inspection, we can reconfirm that the statement "Sirius is white" is credible.

There is another matter of rejoice. Since ancients took the five-element and five-color system as the fixed pattern, they must adopt a flexible attitude toward the marginal colors of the five stars and classify them into the five-color system. In this way, when they talked about star colors, they could be inevitably inaccurate. However, as for the color of Sirius, it is a strife between red and white, both of which are in the five-color system. So, there is no need to worry about similarity, marginality, or flexibility. This further ensures the reliability of using ancient Chinese literature to solve Sirius's color change.

Now, we already know that only ancient records of the five-color standards are credible. However, Sima Qian's five-color standard is only a single evidence, can we find more evidences? After an extensive research of the astrological literature before the seventh century in China (as after the seventh century, there has been no records of Sirius being red in the Western literature), I have found the following four records:

---

[13] Bo Shuren *et al.*, "On the Changes of Betelgeuse (α Ori) in the Past 2000 Years", *Science and Technology History Collection*, 1st edition. (Shanghai Science and Technology Press, 1978), pp. 75–78.

	Text	Source	Author	Time
1	Sirius	*Shih Chi-Tian Guan Shu*	Sima Qian	100 B.C.
2	Sirius	*Han Shu-Tian Wen Zhi*	Ban Gu, Ban Zhao and Ma Xu	100 A.D.
3	Sirius and Vega	Jingzhou Zhan in Vol. 45 of *Kai Yuan Zhan Jing*	Liu Biao	200 A.D.
4	Sirius	*Jin Shu-Tian Wen Zhi*	Li Chunfeng	646 A.D.

As the research given in the columns Author and Time of the above table is boring, I will not include it here.[14] But Item 3 about Jingzhou Zhan, Jingzhou Divination on Sirius and Vega, is worth noting. Vega 1 (Lyra α) is of the same type as Sirius, a white bright star, which further confirms the reliability of the record on Sirius's color.

So, we can safely draw the conclusion that starting from at least 2,000 years ago, Sirius has been seen as the standard white star by ancient Chinese astrologers. Thus, in those reliable ancient records in China, Sirius has always been white not red. This being so, the current theory of stellar evolution shall not be challenged with regard to the color of Sirius.

I published the paper "The Records of Sirius in Chinese Ancient Books" in *Acta Astronomica Sinica* in 1992 and in the second year, the English version of the full text was published in a British journal. Ceragioli, known for his study of Sirius's color issue, made the following comment in the authoritative *Journal for the History of Astronomy*:

> "So far, the best analysis of the Chinese literature published in English was done by Jiang Xiaoyuan in 1993. After extensive literature review, Jiang concluded that there were only four historical records about Sirius's color in the early Chinese literature. And the four records all state that the color of Sirius color is white."[15]

---

[14] See Jiang Xiaoyuan, "The Records of Sirius in Chinese Ancient Books", *Acta Astronomica Sinica*, Vol. 33, Issue 4 (1992).

[15] Ceragioli, R. C., "The Debate Concerning 'Red' Sirius", *Journal for the History of Astronomy*, Vol. 26, Part 3 (1995).

This can be regarded as a result of making the past serve the present. Of course, seen from the perspective of astronomical development, its importance can never match the above-mentioned *A New Catalogue of Ancient Novae*.

I find it necessary to remind our readers that apart from the above two cases of making the past serve the present, there is another direction that makes ancient records of celestial phenomena serve modern astronomy, i.e., ancients used records of eclipses and lunar occultations to study the rotation of the earth. As work in this direction wasn't that extraordinary, it did not produce achievements as great as the *A New Catalogue of Ancient Novae*. Many astronomers have conducted research in this regard, which I will refrain from introducing here. You must still remember that I have made witty remarks on equation in the preface. Without the help of equations, the earth's rotation can hardly be expounded, which, unfortunately, is beyond my ability.

5. Legacy three: The time King Wu conquered King Zhou and the celestial phenomena

The second category of legacies left by Chinese *tianxue* is that *tianxue* can be used to solve issues of astronomical chronology. As some historical documents of the remote past are nowhere to be found, the time of historical events or the birthdays of important historical figures can hardly be ascertained. Historical chronology attempts to solve such problems. Luckily, the theory of Interactions Between Heaven and Mankind was prevalent among ancients, irrespective of whether they are in the Orient or in the West. Ancients believed that there was a kind of mythical connection between heaven and the human world, so in recounting the occurrence of important events or the births and deaths of important figures, ancients would devoutly record the special celestial phenomena of the time, such as solar and lunar eclipses, comets, novae and special locations of planet. Parts of these records have been preserved to the present day. These astronomical records kept for astrological purposes are surprise legacies for historians, which could become persuasive evidences for the determination of historical events if these celestial phenomena could be calculated by the use of modern astronomical expertise.

It can be said that among Chinese historical chronological problems, the most notable and the most fascinating issue is perhaps the chronological times of King Wu's Crusading Against the Tyrant Zhou of the Shang Dynasty (approximately 17th-11th century B.C). Because of the rich and yet uncertain historical data handed down, the time of this event can hardly be ascertained, making the research involving many aspects difficult to collate, such as textual research of historical data, compilation of ancient calendars, the ancient astronomical calculations of ancient celestial phenomena and interpretation of bronze inscriptions and so on. This research project has been most appealing to scholars at home and abroad. It enables them to exercise their textual research capabilities and imagination. It's amazing that this project has been lasting so long, involving so many researchers.

The earliest scholar undertaking this research might be Liu Xin at the end of the Western Han Dynasty. Chapter "Shi Jing" in *Han Shu-Lü Li Zhi* is the historical chronological achievement Liu Xin attained by following *San Tong Li*. We can revere Liu as the father of Chinese historical chronology. In his "Shi Jing", he presumed that King Wu's Crusading Against the Tyrant Zhou took place in 1122 B.C. For a period of 2000 years, this presumption has been agreed upon by many scholars. In *Xin Tang Shu-Lü Li Zhi*, there is a chapter "Da Yan Li Yi" by Yi Xing of the Tang Dynasty, which states that the event of King Wu's Crusading Against the Tyrant Zhou took place in 1111 B.C. This conclusion has been well backed by modern scholars including Dong Zuobin.

Most researchers in the ancient times agreed with Liu's viewpoint regarding the year of King Wu's Crusading Against the Tyrant Zhou, but there were also scholars holding differing views. Since the onset of the 20th century, more researchers have joined the research, including scholars from China, Japan, Europe and the U.S. Many treatises have come out at home and abroad. As of 1997, more than 100 treatises on the subject have been published, among them, 44 have proposed different years of the event, some even with the difference of more than 100 years.[16] Some scholars have kept studying the subject and renewing their conclusions.

---

[16] See Beijing Normal University Institute of Chinese Studies, *Studies on the Year of King Wu's Crusading Against the Tyrant Zhou* (Beijing Normal University Press, 1997).

Amid the wealth of studies on the year of the event, the most famous one was done by Zhang Yuzhe, the late Director of the Purple Mountain Observatory of China. Based on the records of Halley's Comet, Zhang presumed the year of the event. He concluded that "provided the comet recorded in the *Huai Nan Zi.-Bing Lüe Xun* in the record about King Wu's Crusading Against the Tyrant Zhou was truly Halley's Comet, the event must have taken place in 1057 B.C."[17]

Zhang Yuzhe's paper immediately caused repercussions in the academia. As this treatise was published in *Acta Astronomica Sinica*, a journal seldom read by the majority of arts scholars, Zhao Guangxian believed that "it has got a scientific basis, more convincing than previous ones".[18] He went so far as to publish an article in *Historical Research* to add supplementary remarks on it, thus making Zhang's view even more popular. Many humanists chose to believe it, as evidenced by the remarks of Li Chaoyuan, the Deputy Curator of Shanghai Museum. Li said as follows: "The 1057 view is believed to be the most scientific conclusion, which has been well implanted in our mind."[19]

However, the humanistic scholars' high acknowledgement of Zhang Yuzhe's treatise is more than what Zhang expected. Today, while reviewing his article, we cannot help but admire his rigorousness as an astronomer. In his conclusion, Zhang said thus:

"If the comet emerging at the time of King Wu's Crusading Against the Tyrant Zhou was truly the Halley comet, it must have taken place between 1057 and 1056 B.C."

The problem is that it is a hypothetical sentence. It is virtually impossible to determine whether the comet mentioned in *Huai Nan Zi-Bing Lüe Xun* is truly Halley's Comet. My PhD candidate Lu Xianwen examined it in his doctoral dissertation *Research of Records of Comets in Ancient*

---

[17] Zhang Yuzhe, "The Trends of Orbit Evolution of Halley's Comet and Its Ancient History", *Acta Astronomica Sinica*, Vol. 19, Issue 1, (1978).

[18] Zhao Guangxian, "On Deducing The Year of King Wu's Crusading Against the Tyrant Zhou", *Historical Research*, Issue 10, (1979).

[19] Li Chaoyuan, E-mail sent to Jiang Xiaoyuan on January 10, 1998.

China and he came to a very convincing conclusion. Here is a brief account of his research.

For periodic comets, theoretically the tropical year can be deduced by the use of methods of dynamics. But for subjects such as when King Wu's Crusading Against the Tyrant Zhou took place, the controversial range of time periods could be around 100 years. Therefore, it is pointless to turn to comets that come at intervals longer than 200 years and shorter than 20 years.

Only the Halley-type comets (with cycles between 20 and 200 years; note that Halley's comet is a typical comet) can help determine the years. The long-term movement of the Halley-type comet is stable. If we can modify the historical records while pushing back the integral process of the orbits, we can derive fairly reliable results.

Japanese scholar Hasegawa Ichiro made a survey of macroscopic comets during the 200 years from 1700 to 1900.[20] According to his survey, there are 177 appearances of comets that are larger than Grade-6, among which Halley's comet appeared six times, only accounting for 5% of the total. According to the current theory, the number of comets appearing in the last 3000 years is even. Therefore, we can presume that the above proportion also applies to situations of 1000–1100 B.C. In other words, the Halley-type comet only takes about 5% of the total.

Then a comparison of the comet records between 1700 and 1900, and the "Macroscopic Comets Table" of Hasegawa Ichiro finds that official historical records are more reliable than local records. Therefore, we counted the occurrences of comets in official records. From 1 A.D. to 1500 A.D., there are altogether 345 occurrences of comets recorded (multiple records happened to be about the occurrence of the same comet, only observed from different times and different locations and these records were deemed as one occurrence). There are so far 21 occurrences of Harley's comets, which have been confirmed as short-period comets, accounting for 6% (of which there are 19 records of the Halley's comet itself). This result is consistent with what Hasegawa Ichiro derives from his research into Western historical documents.

---

[20] Hasegawa, I., *Vistas in Astronomy*, Vol. 24, (1980), p. 59.

Now, let's apply the above results to the issue of King Wu's Crusading Against the Tyrant Zhou. From the above discussion, we have understood that there were only 5–6%, or even smaller, chances that the comet appearing at the time of King Wu' s Crusading Against the Tyrant Zhou happened to be Halley comet. On the other hand, a total of 23 Halley-type comets have been identified, of which 17 appear at intervals of less than 100 years. In other words, a back-casting finds that at least 17 comets appeared between 1000 and 1100 B.C. Therefore, the probability to determine one of them is 1/17. If we multiply 1/17 with 5–6%, we can get the probability of the comet in *Huai Nan Zi-Bing Lüe Xun* being the Halley's comet, which is 0.0029-0.0035. Simply put, it's only about 0.3%. Common sense tells us that the conclusion cannot obviously be based on the probability of 0.3%.[21]

Let's take a step back. Even if we knew what specific Halley-type comet it is, it would still be difficult to determine the exact time in 1000–1100 B.C. For example, the return of Halley's comet has been confirmed more than 30 times as the orbit parameters of Halley-type comet is most precisely known. While trying to get the exact time of the comet's recurring near perihelion between 1100 and 1000 B.C., scholars obtained different results, with the discrepancy of over two years. These results could never be consistent with the observation records. For example, from the comet ephemeris, we can see that Zhang Yuzhe's calculation is consistent with the records of the comet appearing at the time when King Wu's Crusading Against the Tyrant Zhou took place. The calculations of Yeomans and Kiang (Jiang Tao) are not consistent with the record. In other words, the comet appearing at that time was not likely to be the Halley's comet, and it is less likely to be some other comet as the discrepancies would be larger. Therefore, we can draw the following conclusion: We cannot determine the year of King Wu's Crusading Against the Tyrant Zhou based on the existing comet records.

---

[21] Here, it refers to visual magnitude, indicating the degree of brightness observed with naked eyes. The larger the number, the darker the star. Usually, the lower limit of being visible to the naked eye is 6. The brightest stars could be negative, for instance, Sirius is −1.6 while the sun is −26.74.

This may sound discouraging, but this seems to be the case. Scientific research is to pursue the truth, which shall not be hampered even if it goes against our best wishes.

Fortunately, Heaven will always leave a door open. There are far more than one astronomical record during the time King Wu started crusading against the tyrant Zhou. In fact, there are more than 10 historical records in ancient Chinese books, involving more than 10 kinds of astronomical phenomena. But some of them are meaningless, outlining celestial phenomena that could never have happened during the controversial years of King Wu's Crusading Against Zhou. Some records are too broad and vague. According to our study, there exist the following records, which might help determine the year.

1. *Ligui inscriptions:* King Wu started to crusade against the tyrant Zhou on the morning of a Jiazai day. The divination says the tyrant King Zhou was fatuous and the war would defeat King Zhou.

2. *Shangshu-Zhoushu-Wucheng quoted in Han Shu-Lü Li Zhi* (2): On the Renchen day of the first month when the moon was at the last quarter (the next day being Guisi), King Wu departed from Zhou to crusade against the tyrant King Zhou.

   On the Hui day, the last day of the second lunar month, five days after Jiazi, King Wu killed King Zhou; on the day of the fourth month (when the moon was at the last quarter), namely six days after Gengxu, King Zhou held a sacrifice ceremony at the ancestral temple of the Zhou royals; on the following Xinhai day, a sacrifice ceremony was held; on Yimao five days later, King Wu presented captives to the ancestral temple.

3. *Yi Zhou Shu Vol. 4 — Shi Fu Jie 40*: On the Renchen day of the first month when the moon was at the last quarter (the next day being Guisi), King Wu departed from Zhou to crusade against the tyrant King Zhou. In the last quarter of the moon of the second month, a Jiazi day five days later, the troops arrived at the battlefield and confronted the Shang troops. They then killed King Zhou.

4. Remarks of Ling Zhoujiu to King Jing of Zhou recorded in *Guo Yu-Zhou Yu* (2): Previously, King Wu started to crusade against King Zhou on the Day Ximu of Month Tiansi in the Year Chunhuo at the

hour Doubing when the star was at Tianyuan. The star and the time locations were all in the north.

5. *Huai Nan Zi-Bing Lüe Xun*: Wu Wang Fa Zhou (when King Wu started to crusade against the tyrant King Zhou), the Jupiter was in the east.

These astronomical phenomena are about the locations of the sun and the moon, planetary position and phases of the moon. Among them, the moon-phase terms in "Wu Cheng" and "Shi Fu" have been puzzling scholars, such as "Pang Si Ba", "Ji Si Ba" and "Ji Pang Sheng Ba". Liu Xin was the first person to propose the "fixed-point" theory, which was used to interpret the chapters of "Wu Cheng" and "Shi Fu". But it was problematic when it was used to interpret the inscriptions of unearthed bronze wares. Based on Li Xueqin's latest research results,[22] we turn to the fixed-point theory to interpret the moon-phase terminologies in "Wu Cheng" and "Shi Fu". Based on these celestial phenomena, together with my doctorate students, Ding Xian and Lu Xianwen, I've worked out the timetable of King Wu's crusading against King Zhou, by the use of the most advanced ephemeris software DE404, as follows:

The ultimate conclusion is that King Wu sent out troops on December 4, 1045 B.C. and waged the Muye Battle on January 9, 1044 B.C. He finally overthrew the Shang Dynasty.

6. What is the most important legacy?

After the two case studies above, we are coming to the close of the chapter. In fact, be it to solve modern astrological problems or historical chrono-logical issues, we make use of the small portion of the legacies of Chinese *tianxue*. Then how can we make use of the major part of Chinese *tianxue*? Frankly speaking, Chinese people have suffered a lot from pragmaticism (not sure how it's of use in any way). We should tolerate what is not useful. Unfortunately, I'm also one of the victims. If you insist on the usefulness, I would say with *tianxue*, we can learn about Chinese ancient society.

---

[22] Li Xueqin, "Moon Phases Recorded in *Shang Shu* and *Yi Zhou Shu*", *Chinese Culture Research*, Summer Edition, Issue 2, (1998).

Timetable of King Wu's Crusading Against the Tyrant Zhou (1045–1044 B.C.)

Date (B.C.)	Heavenly Stems and Earthly Branches	Mercury Longitude	Jupiter Longitude	Moon Longitude	Sun Longitude	Calendar Day	Important Records
1045.11.7.	Xinyou	208.63	164.09	209.55	215.63	Twelfth month 1	**New moon**
1045.12.3.	Dinghai	250.32	167.95	193.72	242.19	27	**King Wu started to crusade against King Zhou on the Day Ximu of Month Tiansi in the Year Chunhuo**
1045.12.4.	Wuzi	251.98	168.07	206.30	243.21	28	**Sending out army**
1045.12.7.	Xinmao	256.99	168.42	243.08	246.27	First month 1	**New moon (First month in Shang, second month in Zhou)**
1045.12.8.	Renchen	258.66	168.53	255.06	247.29	2	*On the Renchen day of the first month when the moon was at the last quarter*
1045.12.9.	Guisi	260.33	168.64	266.94	248.31	3	*King Wu departed from Zhou to crusade against the tyrant King Zhou*
1045.12.22.	Bingwu	279.90	169.82	76.24	261.55	16	**Full moon:** on the Renchen day of the first month when the moon was at the last quarter
1045.12.23.	Dingwei	280.99	169.89	91.51	262.57	17	King Wu departed from Zhou to crusade against the tyrant King Zhou

(Continued)

(*Continued*)

Date (B.C.)	Heavenly Stems and Earthly Branches	Mercury Longitude	Jupiter Longitude	Moon Longitude	Sun Longitude	Calendar Day	Important Records
1044.1.5.	Gengshen	281.34	170.55	263.88	275.75	30	*In the last quarter of the moon of the second month*
1044.1.6.	Xinyou	280.30	170.58	275.69	276.76	Second month 1	**New moon (Second month in Shang and third month in Zhou)**
1044.1.9.	Jiazi	276.89	170.65	311.17	279.80	4	**Conquering Shang:** *They then killed King Zhou*
1044.1.12.	Dingmao	273.66	170.70	347.63	282.82	7	Jiang Taigong ordered Fanglai to present captives to the ancestral temple of Zhou on the Dingmao day
1044.1.13.	Wuchen	272.73	170.70	0.25	283.83	8	King Wu then burned incense to offer sacrifice to Heaven and King Wen of Zhou and on the very day, he appointed commander of the army
1044.1.17.	Renshen	270.18	170.70	54.63	287.86	12	Lü Ta was ordered to crusade against Xi Fang and on the 13th day (Renshen), Lü Ta returned to report to King Wu the number of enemy troops killed and the number of captives

1044.1.26	Xinsi	271.11	170.52	186.60	296.88	21	Hou Lai was ordered to crusade against the enemy and gather the army in Chen and on the 22nd day (Xinsi), he returned to report to King Wu the number of enemy troops killed and the number of captives
1044.1.29	Jiashen	273.03	170.40	224.88	299.88	24	Bai Yan vowed to fight against Wei and he returned to report to King Wu the number of enemy troops killed and the number of captives
1044.2.4.	Gengyin	278.43	170.09	296.26	305.86	Fourth month 1	*New Moon (revised)*
1044.2.19	Yisi	297.84	168.87	137.73	320.73	16	**Full moon:** *on the day of the fourth month (when the moon was at the last quarter)*
1044.2.24	Gengxu	305.66	168.35	208.02	325.65	21	*King Zhou held a sacrifice ceremony at the ancestral temple of the Zhou royals*
1044.2.25	Xinhai	307.29	168.24	220.86	326.63	22	Sacrificial ceremony: King Wu offered the Nine Tripod Cauldrons captured from the Shang. He made known the news to Providence, gods, deities and the ancestors.

(Continued)

(Continued)

Date (B.C.)	Heavenly Stems and Earthly Branches	Mercury Longitude	Jupiter Longitude	Moon Longitude	Sun Longitude	Calendar Day	Important Records
1044.2.26	Renzi	308.95	168.13	233.33	327.61	23	Without changing his sacrifice clothes, he came to the ancestral temple with a huge yellow axe
1044.2.27	Guichou	310.64	168.01	245.51	328.60	24	King Wu sacrificed about 100 captured nobles from the Shang
1044.2.28	Jiayin	312.35	167.90	257.47	329.58	25	King Wu told the Wuye victory over the Shang to his forefathers
1044.3.1	Yimao	314.08	167.78	269.31	330.56	26	*King Wu led dukes and princes to the ancestral temple for a sacrifice ceremony*
1044.3.6	Gengshen	323.12	167.19	329.04	335.45	Fifth month 1	*New moon*

Note: In the rightmost column, those in boldface are descriptions, those in italics are from Wucheng while those in regular script are from Shifu.

If you have read the book from the very beginning, you would have been informed that ancient China did not have astronomy in the modern sense. Ancient China only had *tianxue*, which was not a natural science, but a kind of spiritual life of ancient Chinese. The occurrences of solar eclipses, special positions of Venus or Jupiter, and the shooting of a comet were not scientific in their eyes (they had heard of the word *science*), but philosophical, theological or even political, though they had never heard of the word *politics*.

Due to the fact that *tianxue* enjoyed such a privileged status that could never be matched by disciplines, such as mathematics, physics, alchemy, textile, medicine, and agriculture, *tianxue* becomes an important vehicle for us to learn about the political, spiritual and social lives of ancient Chinese. All *tianxue*-related literature passed down from ancient China is useful in this regard. For specific cases, readers are directed to consult the first two chapters of my book *Tianxue Zhenyuan* (*The Truth of the Sciences of the Heaven*). With historical research getting deeper, particularly with the introduction of sociological and cultural anthropological approaches, Chinese *tianxue* is sure to promise more.

Postscript

I have spent two years writing this book. I'm grateful to Ms. Luo Xiang, editor of Shanghai People's Publishing House for her patience, as she has been encouraging me to finish it. Now, I have finally finished it. In retrospect, I have some lingering fear. Upon turning 40, I have felt I am increasingly occupied by chores. My time is mercilessly segmented by various obligations and desires. I have been thinking when I can put all my heart into reading and writing. It has been 10 years since I was last obsessed with reading and writing. I thought I could have a few days totally at my disposal. But, before the postscript is even finished, there are several tasks awaiting me. So I have to accept what is predetermined for me.

I would like to pay special thanks to my two doctorate students, Niu Weixing and Lu Xianwen, whose research work has benefited me a lot. I must also thank my daughter Jiang Tianyi. As a trade for my permission for her to play PC games, she has written some words in some chapters of this book, which proved to be good, though the motive behind the penning of these words is not.

Jiang Xiaoyuan
Shuangxitang, Shanghai
July 16, 1998

When I received the proof sheets of the book, I was transferred to Shanghai Jiaotong University, where I act as the dean of the Department of History of Science and Philosophy, the first of its kind of China. The Department was jointly established by the Institute of Natural Sciences from the Chinese Academy of Sciences (CAS) and the Institute of Humanities and Social Sciences of Shanghai Jiaotong University, which is a landmark event in the history of sciences as it has been made into a discipline in China.

Jiang Xiaoyuan
March, 1999

Index

303

Printed in the United States
by Baker & Taylor Publisher Services

Printed in the United States
by Baker & Taylor Publisher Services